Monoclonal Antibodies

Edited by Nima Rezaei

Published in London, United Kingdom

IntechOpen

Supporting open minds since 2005

Monoclonal Antibodies
http://dx.doi.org/10.5772/intechopen.77874
Edited by Nima Rezaei

Contributors
Shridhar Ghagane, Maya Kulshekar, Sridevi Puranik, Rajendra Nerli, Murigendra Hiremath, Jane Zveiter Moraes, Barbara Hamaguchi, Camila Braggion, Enzo Speciale, Fernanda Cesar, Gabriela Soares, Juliana Osaki, Rodrigo Aguiar, Tauane Pereira, Mepur Ravindranath, Fatiha E.L. Hilali, Hanley Abramson, Ana M. Maria Moro, Tania Maria Manieri, Carolina Georg Magalhães, Daniela Y. Yumi Takata, João Victor Batalha-Carvalho, Atis Chakrabarti, Jukka Kervinen, Egbert Müller, Toru Tanaka, Kazuaki Muranaka, Nima Rezaei, Mona Sadeghalvad

First published in London, United Kingdom, 2021 by IntechOpen
IntechOpen is the global imprint of INTECHOPEN LIMITED, registered in England and Wales, registration number: 11086078, 5 Princes Gate Court, London, SW7 2QJ, United Kingdom
Printed in Croatia

British Library Cataloguing-in-Publication Data
A catalogue record for this book is available from the British Library

Additional hard and PDF copies can be obtained from orders@intechopen.com

Monoclonal Antibodies
Edited by Nima Rezaei
p. cm.
Print ISBN 978-1-83968-369-5
Online ISBN 978-1-83968-370-1
eBook (PDF) ISBN 978-1-83968-371-8

For EU product safety concerns:
IN TECH d.o.o., Prolaz Marije Krucifikse Kozulić 3, 51000 Rijeka, Croatia, info@intechopen.com or visit our website at intechopen.com.

Meet the editor

Professor Nima Rezaei obtained an MD from Tehran University of Medical Sciences, Iran, and an MSc in Molecular and Genetic Medicine and a Ph.D. in Clinical Immunology and Human Genetics from the University of Sheffield, UK. He also completed a short-term fellowship in Pediatric Clinical Immunology and Bone Marrow Transplantation in the Newcastle General Hospital, England. Dr. Rezaei is now Full Professor of Immunology and Vice Dean of International Affairs, School of Medicine, Tehran University of Medical Sciences, and the co-founder and Deputy President of the Research Center for Immunodeficiencies. He is also the founding president of the Universal Scientific Education and Research Network (USERN). Dr. Rezaei has been the director of more than fifty research projects and has designed and participated in several international collaborative projects. He is also an editorial assistant or board member for more than thirty international journals. He has edited more than 30 international books, has presented more than 500 lectures/posters in congresses/meetings, and has published more than 900 articles in international scientific journals.

Contents

Preface

Immunology science has made prominent progress in recent years, helping in all medical fields from diagnosis to treatment. Nowadays, immunotherapy procedures are used in the treatment of a variety of autoimmune diseases, cancers, and organ transplants. It has also paved the way for medical researchers toward a promising future in the treatment of immunological disorders. Immune-based therapies are widely explored in several immunological disorders due to their high specificity and sensitivity. Since the approval of the first monoclonal antibody (mAb) in 1986, mAbs have been used as a novel way of targeting antigens in these disorders. Currently, mAbs are an important group of therapeutic molecules in clinical trials for treating disorders such as inflammatory and autoimmune diseases, malignancies, and cardiovascular and infectious diseases.

This book provides knowledge about several types of mAbs and their application in the clinic and laboratory. The introduction describes mAbs, their structure, and their production process. Subsequent chapters cover more information about the therapeutic application of mAbs, especially in urology and oncology and for multiple myeloma. Furthermore, the book discusses the production and characterization of mAbs. In this regard, chapters examine alternative methods to animal use for generating mAbs as well as novel analytical and *in silico* techniques for characterizing them. These methods can provide comprehensive information about technologies associated with mAb production.

The book describes and discusses immunology concepts and mechanisms related to the function of mAbs in a clear and simple way, making the information herein useful for scientists and clinicians in various fields.

Nima Rezaei, MD, Ph.D.
Research Center for Immunodeficiencies,
Children's Medical Center,
Tehran University of Medical Sciences,
Tehran, Iran

Network of Immunity in Infection,
Malignancy and Autoimmunity (NIIMA),
Universal Scientific Education and Research Network (USERN),
Tehran, Iran

Chapter 1

Introduction on Monoclonal Antibodies

Mona Sadeghalvad and Nima Rezaei

Abstract

Monoclonal antibodies (mAbs) are a group of antibodies produced by identical clones of B lymphocytes against a particular antigen. mAbs are identical in several properties such as protein sequence, antigen-binding site region, binding affinity for their targets, and identical downstream functional effects. These characteristics of mAbs highlight their differences with the polyclonal antibodies which have heterogenous activities and recognize different epitopes on an antigen. Murine mAbs was the first generation of mAbs developed by hybridoma technology however, because of their murine origin, they can trigger the anti-mouse antibody response in the host which could accelerate mAb clearance and undesirable allergic reactions upon repeated administration. This issue was resolved by developing engineering methods toward producing less immunologic chimeric or humanized antibodies. mAbs applications have become a novel way of targeting antigens in a wide variety of diseases such as autoimmunity, malignancies, and asthma. In addition, high specificity and high affinity binding properties of mAbs make them effective biological reagents in immunodiagnostic assays. They can be used in diagnosis of infectious diseases and detection of certain antigens or in serological assessments for detection of antibodies against a certain antigen. This chapter summarizes the general properties of mAbs, their production processes, and their important diagnostic and therapeutic applications.

Keywords: monoclonal antibodies, mAb, chimeric mAb, humanized mAb, fully humanized mAb

1. Introduction

Antibodies or immunoglobulins (Ig) are glycoproteins produced by differentiated B lymphocytes named "plasma cells" in response to exposure to antigens. The diversity of antibody responses to different antigens is because of the gene recombination process in the hyper-variable regions of antibodies. During the recombination process in their genes, antibodies undergo gene rearrangement that allows them for diverse binding [1]. High specificity and diversity of antibodies have made them popular molecules with very high efficiencies in several therapeutic or diagnostic applications.

Monoclonal antibodies (mAbs) are a group of antibodies produced by identical clones of B lymphocytes against a particular antigen. Monoclonal antibodies are identical in several properties such as protein sequence, antigen-binding site region, binding affinity for their targets, and identical downstream functional effects. These characteristics of mAbs highlight their differences with the

Type of mAb	Description	Structure
Murine mAb	Murine mAbs was the first generation of monoclonal antibodies developed by hybridoma technology. They have no human components in their structure and could result in producing the human anti-mouse antibodies (HAMAs). Suffix: -Omab e.g.: Abagovomab (anti CA-125 in ovarian cancer)	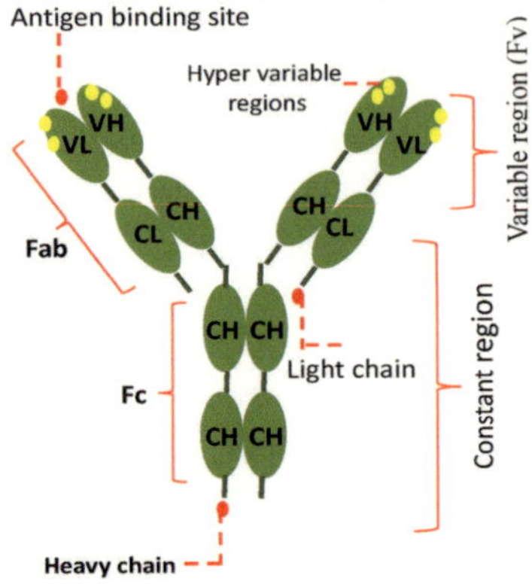
Chimeric mAb	In chimeric mAb, constant regions are humanized but variable regions in both heavy and light chains remain murine Suffix: -Ximab e.g.: Rituximab, Infliximab	
Humanized mAb	Hyper variable regions are murine Suffix: -Zumab e.g.: Natalizumab, Gemtuzumab	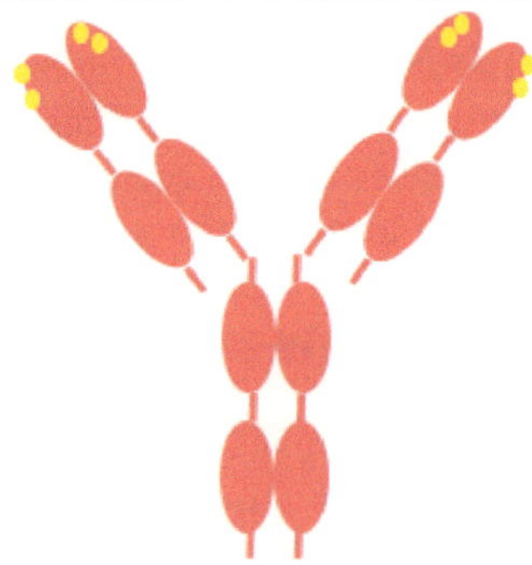
Fully human mAb	100% human Suffix: -Umab e.g.: Ibritumab, Ofatumumab	

Table 1.
Different types of monoclonal antibodies. Murine mAbs were the first generation of mAbs with higher immunogenicity in humans. Gene engineering methods provide the less immunogenic mAbs by replacing human components in mAb structure. ***mAb:*** *Monoclonal antibody.*

polyclonal antibodies which have heterogenous activities and recognize different epitopes on an antigen.

Using mAbs has become a novel way of targeting antigens in a wide variety of diseases and conditions since the first mAb was approved in 1986. Orthoclone OKT3® (muromonab-CD3) was the first mAb approved by the Food and Drug Administration (FDA). OKT3 was produced based on murine hybridoma technology by Kohler and Milston for the treatment of acute transplant rejection [2]. Currently, mAbs are the important group of therapeutic molecules in clinical trials for treating different disorders such as inflammatory and autoimmune diseases (e.g. rheumatoid arthritis, systemic lupus erythematosus, psoriasis, inflammatory bowel diseases), malignancies (e.g. leukemia, melanoma, breast cancer, and multiple myeloma), cardiovascular, and infectious diseases [3].

Murine mAbs was the first generation of monoclonal antibodies developed by hybridoma technology. They have no human components in their structure and could result in producing human anti-mouse antibodies (HAMAs). HAMA response caused hypersensitivity reactions (e.g. anaphylaxis and serum sickness) in the recipients, resulting in fast clearance of antibodies or reducing their effectiveness [4]. Genetic engineering approaches and using transgenic animals were developed to overcome these troubles; So that a transformed cell line could produce the altered antibody structurally closer to human antibodies. These modified antibodies are known as chimeric mAbs because their constant region is human while their variable region is murine (**Table 1**). This technology was developed for the first time in 1980s by scientists in Cambridge, UK. After that, humanized and fully human mAbs were developed to reduce mAb immunogenicity and their side effects. Humanized antibodies have human light and heavy chains but hypervariable regions are still murine while fully human antibodies are totally humanized. However, they are still immunogens and may have important adverse effects caused by production of antidrug antibodies (ADAs) [5]. This chapter summarizes the general properties of mAbs, their production processes, and their important applications, including therapeutic and diagnostic uses.

2. Antibody structure and functions: immunoglobulin G as the therapeutic mAb

An antibody molecule has a Y-shaped structure with a total molecular weight of ~150 kDa, composed of four polypeptide chains including two identical heavy (H) and two light (L) chains (**Figure 1**). Covalent bonds (mainly disulfide interactions) provide the stability of heavy and light chains next to each other. Each heavy or light chain is composed of constant (CH and CL, respectively) and variable domains (VH and VL, respectively) [4].

Each antibody has two identical arms known as "antigen binding fragments" or Fabs, acting as antigen-binding sites. Each Fab consists of a variable region known as Fv (formed by the VH and VL domains), and the constant region (formed by the CH and CL domains). Fv is a highly variable region and responsible for specific binding of antibody to the antigen, contributing to direct effects of antibody such as inhibiting or neutralizing the antigen. There are three hyper variable regions known as complementarity determining regions or CDR1, CDR2, and CDR3 in the variable regions of light and heavy chains, allowing diverse antigenic specificities to be recognized [4]. The Y structure's stem, known as the "fragment crystallizable region" or Fc, is a constant region of the antibody molecule. The Fc region determines the class of the antibody and its functional properties. There are five classes

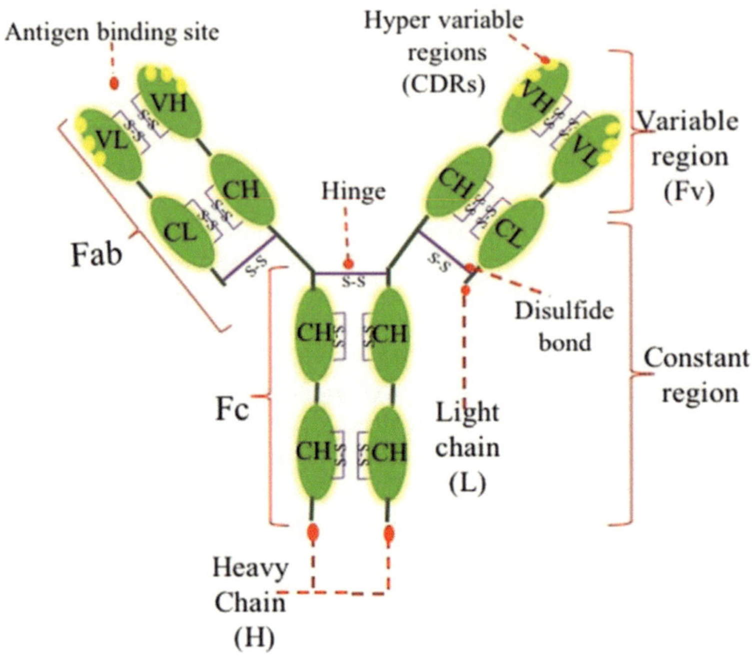

Figure 1.
The schematic structure of an antibody. An antibody molecule is composed of four polypeptide chains including two identical heavy (H) and two light (L) chains. Each heavy or light chain is composed of constant (CH and CL, respectively) and variable domains (VH and VL, respectively). Variable domains form the antigen binding site. CDR: complementarity determining regions; S-S: disulfide bond; C: constant; V: variable.

of antibodies including immunoglobulin G (IgG), IgM, IgD, IgE, and IgA with distinct effector mechanisms for recognition and elimination of the antigens. In addition, the Fc region can interact with a variety of receptors such as Fc receptors or FcRs (expressed on the immune cells) and the components of the complement system (such as C1q). Fc recognition by the immune system components results in initiating the effector functions of antibodies such as antibody-dependent cell cytotoxicity (ADCC), antibody-dependent cellular phagocytosis (ADCP) and complement-dependent cytotoxicity (CDC) (**Figure 2**) [4].

Therefore, several functions are explained by which antibodies can eliminate a particular antigen and both variable and constant regions of antibodies contribute to this response. The stability and flexibility of antibodies and their effector functions such as activating ADCC, CDC, as well as interaction with C1q are important factors determining the suitability of immunoglobulins for the development of therapeutic mAbs. The majority of the clinically available mAbs are IgG. IgG is a glycoprotein with a size of 150 kDa consisting of two heavy and two light chains as described before. A conserved glycosylation site is present at amino acid Arginine297 (N297) in the CH2 domain, playing an important role in the structural conformation of the Fc and its binding to FcRs and complement component C1q [6].

Totally, IgG consists of four subclasses of IgG1, IgG2, IgG3 and IgG4 which differ in their heavy constant region (CH), as well as the hinge structure (the region where Fabs are bound to the Fc region). The difference between hinge regions confers many of the unique characteristics to each IgG subclass, including flexibility, stability and distances between the two Fabs. In addition, the amino acid differences between the binding sites of each subclass could explain the differences in the

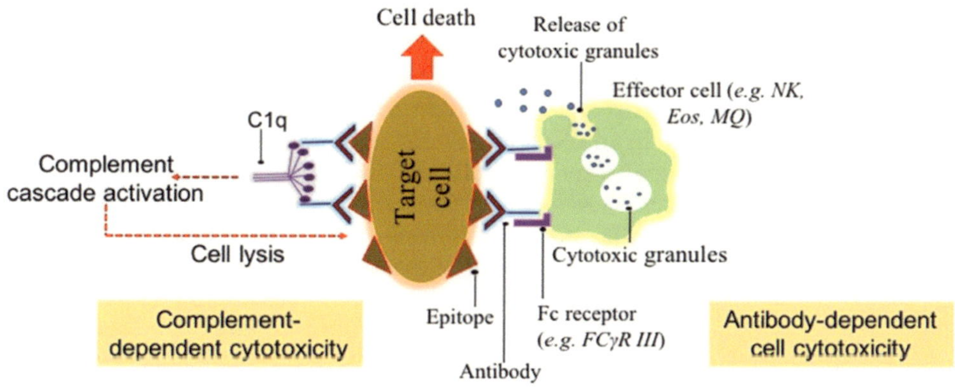

Figure 2.
Two important effector functions of antibody. ADCC is an extracellular killing mechanism leading to antigen elimination. IgG has a bifunctional structure related to the fragment antigen-binding (fab) and fc portions of antibody. ADCC is initiated by the engagement of fab with the antigen from one side, and fc interaction with FcγR on effector cells, from another site. Subsequently, degranulation of effector cells (mainly NK cells) leads to target cell lysis. ***NK:*** *natural killer cell;* ***MQ:*** *macrophage;* ***Eos:*** *eosinophil;* ***ADCC:*** *antibody-dependent cell cytotoxicity.*

effector functions of the IgG subclasses. These variations between IgG subclasses correlate with their selection for therapeutic purposes. Of the IgG subclasses, IgG3 has a longer hinge region compared with other subclasses, making them inappropriate for target binding. On the other hand, IgG3 cannot be purified with protein A and also has the shortest half-life (approximately 7 days) and high allotypic polymorphism compared with other subclasses. So, engineering techniques are required for modifications to the amino acid content of the IgG3 hinge region for development of therapeutics purposes. Meanwhile, most of the mAb therapeutics on the market are composed of IgG1, IgG2 or IgG4 with slow clearance and long half-life properties [6, 7].

IgG1 has high stability and exhibits potent effector functions including ADCC, CDC, and C1q binding being the majority of therapeutic mAbs. IgG1 has the higher affinity for the FcRs compared with the other subclasses (the affinity for Fc receptor: IgG1 > IgG3 > IgG4 > IgG2 respectively) [6].

IgG2 has low affinity for interaction with antigen and also exhibits reduced functional activity compared to IgG1. IgG2 antibodies have three isoforms (known as IgG2-A, IgG2-A/B, and IgG2-B) based on types of disulfide bonds between the antibody chains. These isoforms could be converted to each other. This phenomenon, which is referred to disulfide shuffling, could regulate the activity of IgG2 in the serum [8, 9].

IgG4 has a low affinity for C1q and therefore, this subclass of IgG could emerge as a therapeutic mAb when the host effector function is not desirable. In addition, the exchange of Fab arm is a normal biological process that can occur in IgG4 and is not desirable due to its adverse effects. Natalizumab (Tysabri) and gemtuzumab ozogamicin (Mylotarg) are the examples of therapeutic IgG4 for multiple sclerosis (MS) and acute myeloid leukemia (AML), respectively [6].

3. The production process of monoclonal antibodies

In the following section we described two techniques, including hybridoma and phage display used for the production of mAbs.

3.1 Hybridoma technique

Monoclonal antibodies are generated from a single B lymphocyte clone and bind to the same epitope of an antigen. The hybridoma technique was first used in 1975 to generate mAbs by Milstein and Köhler. Several steps are involved in this method. First, mice are immunized with specific antigens emulsified with appropriate adjuvant. The booster injection is normally done after two weeks and the animal is then sacrificed when enough amount of antibody is produced. Blood collection is performed to assay the sufficient amount of the antibody production using techniques including ELISA and flow cytometry. After sacrificing, the spleen is isolated and then tissue digestion could be applied with an enzymatic or mechanical method leading to release of B cells. B cells could be extracted using density gradient centrifugation [8].

The next step is making a fusion between B lymphocytes and myeloma cells (that are immortal like cancer cells). Prior to fusion, myeloma cells should be prepared by culturing with 8 – azaguanine, making them sensitive to hypoxanthine-aminopterin-thymidin (HAT) medium. The fusion process is carried through using polyethylene glycol (PEG), resulting in cell membrane fusing. After the fusing process, there will be a variety of cells including fused B cells with myeloma cells, unfused B cells, unfused myeloma cells, B cells fused to B cells, myeloma cells fused to myeloma cells. Therefore, a selective medium known as hypoxanthine, aminopterin and thymidine (HAT) medium should be used to select only the B cells fused with myeloma cells [10]. Two components of this medium, hypoxanthine and thymidine, are the metabolites of the salvage pathway of nucleoside synthesis. Therefore, only the cells that have the necessary enzyme for the salvage synthesis of nucleic acids, named hypoxanthine-guanine-phosphoribosyl transferase (HGPRT), will be able to survive. Unfused myeloma cells lack HGPRT, so they cannot replicate their nucleic acid and they will not be able to grow in HAT medium. On the other hand, unfused B cells have a limited life span and therefore cannot grow appropriately. Consequently, only fused B cell-myeloma cells known as "hybridomas" are able to grow in the medium. It should be noted that another pathway of nucleic acid synthesis named "de novo" pathway, is also inhibited due to the presence of aminopterin in HAT medium. So, only the HGPRT-positive cells could be grown in this selective medium [8, 10].

To separate antibodies with different specificity and also for further hybridoma growth, the mixture of cells is diluted in microtiter wells in which their walls are coated with murine macrophages or feeder fibrocyte cells providing the growth factors needed for antibody-producing cells. Then, the antigen-binding ability of secreted antibodies by different clones of B cells could be assessed by ELISA, antigen microarray assay, radio-immuno assay (RIA), or immune-dot blot and finally, the stable clone will be selected. The fused hybridomas and produced mAbs can be stored away in liquid nitrogen [8].

Although this process may be well suited for development of therapeutic antibodies, however, there are some important problems with using this technique. The hybridoma process takes approximately between 6 and 8 months to obtain a sufficient amount of mAbs, so its development procedure is very long. On the other hand, because of the murine origin of the antibodies, they can trigger the HAMA response in the host which could accelerate mAb clearance and undesirable allergic reactions upon repeated administration. This issue was resolved by developing antibody engineering methods toward producing less immunologic chimeric or humanized antibodies. These engineered antibodies were created using murine variable regions or CDRs as well as human constant regions aiming to decrease HAMA response and maintain target specificity. Currently, fully humanized antibodies are generated in transgenic mice models (e.g. HuMabMouse and XenoMouse) using hybridoma technology. For this purpose, the mouse immunoglobulin gene loci have been replaced with human loci within the transgenic mouse genome [8].

3.2 Phage display technique

The phage display method is one alternative to traditional hybridoma technology for generating monoclonal antibodies. This method was developed in 1985 by George P. Smith, who demonstrated that a peptide of interest could be displayed on the surface of filamentous phages following inserting the DNA fragment into the coat protein gene of phage. Then, a process known as "panning or biopanning" is explained by Parmley and Smith; the process describing a selection and affinity enrichment in order to isolation of peptide-phage fusions based on their specific binding affinity. Finally, phage display technology was used for the first time by McCafferty and Winter for generating antigen specific mAbs by creating combinatorial antibody libraries on filamentous phages [11].

This method involves integration of a gene sequence coding for a particular antibody into the DNA sequence of a filamentous bacteriophage leading to the expression of interest protein on the surface of the bacteriophage capsid. These phage libraries could be generated from healthy donors (creating Naïve libraries) or individuals who carry a particular disease, such as metastatic cancer or particular infection, or have been immunized with a particular antigen (creating immunized libraries). M13 is a filamentous bacteriophage that is widely used for antibody production via phage display. This phage infects *Escherichia coli* (*E. coli*) strains.

The discovery of smaller recombinant antibody fragments such as Fv (variable region consisting of VH and VL), Fab, single-chain variable domain (scFv), and diabodies (bivalent scFvs) has played an important role in the advancement of antibody phage display technology [11, 12]. Compared to full antibodies, these fragments are more inclined to expression in bacteria. These fragments can be cloned into a bacteriophage (next to the coat protein known as PIII protein) using a vector. Bacteriophages are then used to infect *E. coli* to generate a library containing approximately 10^{10} cells. Later, bacteriophage containing the antibody segments were secreted from *E. coli*. These cells can then be isolated and sequenced. This technology enables fast and large-scale production of antibodies without animal use and it is easy to screen a large diversity of clones. However, it has some drawbacks, such as more expensive costs and more difficult techniques [11, 12].

4. Applications of monoclonal antibodies: therapeutic and diagnostic uses

4.1 Therapeutic applications of mAbs in cancer therapy

Monoclonal antibodies could be designed specifically against a target antigen found on cancer cells. Several therapeutic mAbs have been approved against different cancer types after the discovery of proto-oncogenes and specific tumor antigens [13]. In 1994, an antibody named MAB 17-1A was approved against epithelial cell surface antigen for identification of adenocarcinomas. It was efficient in reducing the mortality and occurrence rate of colorectal cancer [14]. Rituximab, an anti-CD20 chimeric antibody, was approved in 1997 for treating non-Hodgkin B cell lymphoma. Rituximab interacts with CD20 antigen expressed on B cell tumors and then eliminates malignant cells through an effective immune response [15].

Ibritumab (Zevalin®), Obinutuzumab (Gazyva®), and Ofatumumab (Arzera®) are the other mAbs against CD20 antigen [16].

Epidermal growth factor receptor (EGFR) is another antigen molecule expressed on many human cancer cells involved in cancer progression and metastasis. A fully humanized anti-EGFR mAb has been reported to reduce cancer growth in-vitro and in-vivo. Cetuximab (Erbitux®, C225), a chimeric IgG1, binds to EGFR and induces receptor internalization and degradation. This mAb was approved for the treatment of patients with EFFR-expressing metastatic colorectal cancer (mCRC). Panitumumab (Vectibix®, Amgen) is a fully human IgG2 against EGFR used for the treatment of CRC [17, 18]. Necituzumab (Portrazza®), another EGFR-targeting mAb, is a humanized IgG1 indicated for treatment of patients with metastatic squamous non-small cell lung cancer [16].

Another well-known humanized mAb, trastuzumab (Herceptin®), has been approved for the treatment of breast cancer [19]. Herceptin is an IgG1 mAb that binds to HER2 protein expressed on breast tumor cells and can be used to treat breast tumors with overexpression of HER2 (about 30% of breast cancer patients) [20]. Pertuzumab (Perjeta®) and Ado-trastuzumab emtansine (Kadcyla®) are the other humanized IgG1 mAbs targeting HER2 [16].

Immune checkpoint blockade therapy is another antitumor approach. Immune checkpoint molecules such as Programmed Cell Death Protein 1 (PD-1) and cytotoxic T-lymphocyte-associated protein 4 (CTLA-4) expressed on cancer cells and act as inhibitory receptors which result in suppressing immune response against tumor cells. Anti-PD1/PD-L1 therapy has been shown promising results to treat a variety of cancer types such as lung, liver, blood, and skin cancers [21]. Nivolumab (Opdivo®) is a fully human IgG4 mAb against PD-1 approved for the treatment of metastatic melanoma, metastatic squamous non-small cell lung cancer, and metastatic non-squamous non-small cell lung cancer. Pembrolizumab (Keytruda®) is also a humanized IgG4 mAb targeting PD-1 that was approved for melanoma, lung cancer, and lymphoma. CTLA4 is another checkpoint molecule that could be inhibited by a human IgG1 named Ipilimumab (Yervoy®). CTLA4 plays a critical role in inhibition of T cells, especially during the early stages of T cell expansion. Therefore, ipilimumab could improve T cell activation and promote immune response against tumor [16].

Anti-idiotype mAbs have been considered in cancer therapy because they can mimic tumor associated antigens. Idiotype is referred to antigen binding sites in the variable domain of an antibody molecule. Anti-idiotype mAbs could mimic tumor antigens and may be used as alternate antigens or vaccines for immunization against the tumor [22]. ACA125 is a murine anti-idiotype monoclonal antibody that mimics the tumor antigen CA125. ACA125 was shown to induce anti-anti-idiotypic immune response in the numbers of patients with ovarian cancer associated with prolonged survival [22, 23]. Similarly, good results have been shown in patients with advanced CRC receiving murine anti-idiotype mAb that mimics an epitope of carcinoembryonic antigen (CeaVac) [14, 22]. Another anti-idiotype mAb, TriGem, that mimics disialoganglioside GD2 also demonstrated promising results in patients with melanoma [24].

Fusion proteins consisting of the Fv region of a mAb and a bacterial toxin are also considered as another strategy for cancer therapy which is known as "recombinant immunotoxins". The immunotoxins derived from Pseudomonas enterotoxin shown an effective response against solid tumors as well as lymphomas and leukemias [25].

Radioimmunotherapy using mAbs against cancer cells has also been considered as an efficient therapeutic approach. To this end, mAbs could be labeled with radioisotopes such as iodine-131 and yttrium-90 to deliver radioisotopes to target cells. Iodine-131 and Yttrium-90 were used in the treatment of Hodgkin's disease and lymphoma [26]. Using radiolabeled mAbs was also shown in cancer diagnosis using a diagnostic imaging called immunoscintigraphy [27]. Despite the promising results, there are still several obstacles to the mAbs application in cancer therapy, such as specific targeting without affecting normal cells as well as resistance the tumor cells to drugs [28].

4.2 Therapeutic applications of mAbs in the treatment of autoimmune diseases

Immune system activation in autoimmune diseases or after organ transplantation could be potentially suppressed by mAbs. Successful therapeutic applications of mAbs have been shown in several inflammatory conditions such as psoriasis, rheumatoid arthritis (RA), juvenile arthritis, Crohn's disease, and multiple sclerosis, [29].

4.2.1 Anti-TNF monoclonal antibodies

Because of the crucial role in inflammatory responses, TNF-a is considered as an important cytokine involved in pathogenesis of several disorders such as RA, Crohn's disease, and spondyloarthritides and, therefore, anti-TNF agents have become an efficient approach used in treatment for these diseases. Infliximab (Remicade®) is a human chimeric IgG1 anti-TNF antibody that interacts with soluble and transmembrane forms of TNF-a resulting in inhibiting proinflammatory cascade signaling. Binding infliximab to cells expressing TNF led to cell destruction through antibody and CDC [30, 31]. Inhibiting TNF-a could prevent the production of proinflammatory cytokines such as IL-1, IL-6, and IL-8 [32]. Infliximab was used in 1993 to treat patients with persistent RA. In addition to RA, infliximab was approved to treat crohn's disease, psoriasis, psoriatic arthritis, ankylosing spondylitis, and ulcerative colitis [33]. Moreover, infliximab could also induce T lymphocyte apoptosis in Crohn's disease [34]. Remarkable improvement in clinical parameters such as improvement of joint swelling, pain, reducing the level of inflammatory mediators such as C-reactive protein (CRP) were seen in patients with RA after treating with infliximab [31, 33].

In addition to infliximab, several other anti-TNF mAbs have also been approved for treating autoimmune disorders. These monoclonal antibodies include adalimumab, golimumab, and certolizumab [35]. Adalimumab (Humira®), is a fully human IgG1 mAb neutralizing TNF-a and could induce apoptosis in cells expressing TNF [36]. Adalimumab is approved for use in RA, ankylosing spondylitis, psoriatic arthritis, juvenile idiopathic arthritis, Crohn's diseases, ulcerative colitis, and Psoriasis [37]. Golimumab (Simponi®), a fully human mAb, has been approved for RA, ankylosing spondylitis, psoriatic arthritis, ulcerative colitis, and juvenile idiopathic arthritis [38]. Certolizumab (Cimzia®) is a PEGylated Fab fragment approved for the treatment of Crohn's disease, RA, psoriatic arthritis, and ankylosing spondylitis [39].

4.2.2 Anti-IL-1 and anti-IL-1R monoclonal antibodies

The role of the IL-1 family and their receptors are well-known in inducing and regulating inflammation in autoimmune disorders [40]. Promising results have been shown in patients with autoimmune diseases after using anti-IL-1 mAbs such as Canakinumab, or targeting IL-1 receptor such as anakinra [41].

Canakinumab (ACZ885, Ilaris®) is an anti-IL-1β IgG1 mAb neutralizing IL-1β resulting in inhibition of inflammation in patients with autoimmune disease. Canakinumab was first approved in 2009 for treatment of cryopyrin-associated periodic syndrome (CAPS) [42]. Afterward, this mAb was approved for other inflammatory disorders including TNF receptor associated periodic syndrome (TRAPS), mevalonate kinase deficiency (MKD), familial Mediterranean fever (FMF), and hyperimmunoglobulin D syndrome (HIDS).

Anakinra is an antagonist for IL-1RI which prevents the interaction of IL-1α as well as IL-1β to IL-1R1 resulting in reducing inflammatory response and tissue damage. Anakinra is currently approved for the treatment of RA and cryopyrin-associated periodic syndromes [43]. Other anti-IL1 mAbs are also under investigation for clinical use such as Gevokizumab (anti-IL-1β IgG2 mAb), LY2189102 (anti-IL-1β IgG1 mAb), MABp1 (anti-IL-1α IgG1 mAb), and MEDI-8968 (blocking IL-1RI) [41].

4.2.3 Anti-IL-6 and anti-IL-6R monoclonal antibodies

IL-6 is an inflammatory cytokine involved in the initiation or progression of immune responses in several autoimmune diseases such as RA. Tocilizumab or atlizumab (Actemra® or RoActemra®), is a humanized anti-IL-6 receptor mAb and binds to both soluble and membrane-bound IL-6 receptor. Its efficacy is currently being explored in the treatment of RA, systemic juvenile idiopathic arthritis in children, Castleman's disease, systemic lupus erythematosus (SLE), juvenile dermatomyositis (DM), vasculitis, and juvenile scleroderma [44]. Sarilumab (Kevzara®) is another human IgG1 mAb against IL-6 receptor developed for the treatment of RA [43]. Sirukimab, olokizuman, and clazakumab are the inhibitors of IL-6 that are currently under development for treating inflammatory disorders.

4.2.4 Anti-CD20 monoclonal antibodies

CD20 antigen is a phosphoprotein expressed on B lymphocytes involved in B cell proliferation and activation by initiating an intracellular signaling pathway. Targeting CD20 by mAbs induces B cell apoptosis and could inhibit B cell function through antibody-dependent cell mediated cytotoxicity and complement-dependent cytotoxicity. Rituximab (Rituxan®), a chimeric mAb against CD20 antigen, has been first approved for the treatment of lymphomas. Rituximab was approved for treating RA in combination with methotrexate, which could improve symptoms in patients [45]. Also, promising results have been shown in treating other autoimmune diseases including systemic lupus erythematosus [46], dermatomyositis [47], severe autoimmune hemolytic anemia [48, 49], refractory immune thrombocytopenic purpura [50], Wegener's granulomatosis [51]. Ocrelizumab (Ocrevus®) is another humanized anti-CD20 antibody that targets CD20 molecules on B lymphocytes. It was approved for the treatment of the primary progressive form of multiple sclerosis. Ofatumumab, a fully human anti-CD20 antibody, has been shown to be effective and safe in treating patients with autoimmune diseases. Phase II and III trials are ongoing to evaluate the efficiency of ofatumumab in patients with multiple sclerosis and rheumatoid arthritis, respectively [52, 53].

A phase III trial is ongoing for evaluating a novel glycoengineered chimeric anti-CD20 mAb in patients with relapsing forms of multiple sclerosis (RMS). Glycoengineering led to increased affinity for FcγRIIIa receptors and enhanced ADCC [54].

4.2.5 Other monoclonal antibodies for treating autoimmune diseases

There are several mAbs targeting cytokines or their receptors developed to reduce inflammatory response in autoimmune disorders. IL-17 as a major cytokine of Th17 cells, plays crucial roles in immune response against bacterial and fungal infections, as well as in the pathogenesis of autoimmune diseases, importantly in psoriasis [55]. Secukinumab (Cosentyx®), an IgG1 human mAb, binds to IL-17A and is approved for the treatment of psoriasis and ankylosing spondylitis. Another mAb against IL-17A, named Ixekizumab (Taltz®) also neutralizes IL-17 and was developed for the treatment of moderate to severe plaque psoriasis. Brodalumab (Siliq® or Kyntheum®) is a human mAb also approved for plaque psoriasis. It binds to the IL-17 receptor and inhibits the related signaling pathway. Brodalumab also interacts with IL-17 and prevents its binding to IL-17 receptor [56].

In addition to IL17, a monoclonal antibody targeting IL-23 would be a potential treatment for plaque psoriasis. IL-23 is also a key proinflammatory cytokine playing an important role in Th-17 differentiation and activation. Guselkumab (Tremfya®), Risankizumab (SKYRIZI®), and Tildrakizumab (Ilumya®) are IgG1 mAbs targeting IL-23 p19 approved for the treatment of patients with plaque psoriasis [55].

Targeting adhesion molecules which play an important role in leukocyte activation, circulation, and localization to inflammatory sites is also considered as an efficient therapeutic approach in treating autoimmune diseases [57]. Natalizumab (Tysabri®), a humanized mAb against the cell adhesion molecule α4-integrin, was the first mAb approved for treatment of MS. Natalizumab prevents the interaction of α4-integrin with VCAM-1 expressed on endothelial cells, resulting in the inhibition of leukocyte migration to the central nervous system. Natalizumab is also used for treating Crohn's disease [53]. Alemtuzumab (Lemtrada®) is a humanized mAb against CD52 (or COMPATH1) expressed on lymphocyte, monocytes, and dendritic cells. It could destroy CD52-expressing cells by inducing ADCC. Alemtuzumab was approved for the treatment of patients with multiple sclerosis and chronic lymphocytic leukemia as well as immunomodulation in organ transplantation. Vedolizumab (Entyvio®), a humanized mAb against a4b7 integrin, has also been developed for treatment of patients with Crohn's disease and ulcerative colitis [53, 58].

4.3 Therapeutic applications of mAbs in the treatment of graft-versus-host disease

Two monoclonal antibodies including OKT3 (a murine IgG2a antibody against human CD3) and antibodies against IL-2 receptor (CD25) have been approved to reduce allograft rejection [59].

Graft-versus-Host Disease (GVHD) is a complication of bone marrow transplantation causing death in patients. GVHD occurs when alloreactive donor T cells interact with major histocompatibility (MHC) molecules in the host, leading to immune system activation and releasing higher amounts of cytokines [60]. Targeting T cells before their activation could be effective in inhibiting GVHD. The expression of CD25 on T cells is considered as an important step in their alloreactive activation. Therefore, mAb therapy using anti-CD25 monoclonal antibody might inhibit T cells and could be an effective therapeutic agent [61]. However, the production of anti-mouse antibodies and HAMA response in the host could affect the effectiveness of these mAbs.

Using mAbs for treatment of other complications that occurred post transplantation is also shown. Rituximab, an anti CD20 mAb used for treatment of posttransplant lymphoproliferative disorder [62]. In addition, odulimomab, an anti-LFA1 mAb, was shown to have a protective function against ischemia–reperfusion injury after kidney transplants [62, 63]. Another humanized mAb named Daclizumab (Zynbryta®) targets IL-2 receptor and decreases the risk of acute rejection of renal transplant [64].

4.4 Therapeutic applications of mAbs in the treatment of asthma

High serum levels of immunoglobulin E (IgE) plays an important role in the pathogenesis of allergic asthma causing bronchial hyperresponsiveness [65, 66]. New treatment approaches have been developed to manage disease severity in patients with asthma, including using humanized monoclonal antibodies against IgE or cytokines involved in initiation or persistence of asthmatic inflammation. It has been shown that in patients with moderate to severe allergic asthma, administration of recombinant humanized anti-IgE antibody could result in decreasing serum IgE levels as well as asthma symptoms. These antibodies can exert their effects by forming a complex with free IgE resulting in the inhibition of IgE interaction with its receptor expressed on mast cells and basophils [66]. Omalizumab (Xolair®), a humanized mAb, inhibits IgE binding to its receptor (FcεR1) and showed appropriate efficiency in patients with severe asthma [67].

Targeting IL-4, IL-5, IL-13, IL-9 cytokines could also be an effective approach in the treatment of allergic eosinophilic asthma [68]. IL-4 is an important mediator for TH2 cell differentiation which acts by binding to its receptor, IL-4 receptor (IL-4R), expressed on several types of immune cells. Dupilumab (Dupixent®), a monoclonal antibody against IL-4R, was approved for patients with moderate to severe asthma [69]. Targeting IL-5 could be effective in reducing asthmatic symptoms due to its role in the maturation, activation, and maintenance of eosinophils. Mepolizumab (Nucala®), Reslizumab (Cinqair®), and Benralizumab (Fasenra®) are mAbs against IL-5 approved for eosinophilic asthma [68]. Mepolizumab and Benralizumab block the interaction of soluble IL-5 with its receptor on the eosinophils. Benralizumab binds to the IL-5R expressed on the eosinophils and then inhibits the IL5R signaling pathway. Besides, this mAb can lead to eosinophils' apoptosis through interaction with FcγRIIIa expressed on the natural killer cells [70]. IL-13 is a crucial cytokine involved in IgE production from B lymphocytes causing smooth muscle contractility in asthma [71]. Lebrikizumab and Tralokinumab are mAbs against IL-13 acting by neutralizing IL-13 and inhibiting IL-13 binding to its receptor [72–76]. Targeting IL-9 could be effective in inhibiting mast cell activation. MEDI-528, a humanized IgG1 monoclonal antibody, targets IL-9 and inhibits its function in asthma pathogenesis [77]. Other mAbs such as Tezepelumab (anti Targeting thymic stromal lymphopoietin or TSLP) and Daclizumab (anti IL-2R α chain (CD25)) are also effective in inhibiting the induction of type 2 cytokines (e.g. IL-5, IL-4 and IL-13) and inhibiting lymphocyte activation, respectively [78, 79].

4.5 Other therapeutic applications: using mAbs in the treatment of sepsis and viral infections

Sepsis is considered as an inflammatory immune response and potentially life-threatening disorder that occurs in response to an infection. Bacterial infections are the important cause of sepsis, but other infections including viral, fungal or protozoan infections can also trigger sepsis [80]. Targeting inflammatory mediators such as TNF-α or its receptor could be efficient against inflammatory response. However, inhibiting bacterial toxin or important bacterial components such as endotoxin or lipid-A (gram-negative bacteria component) may be more effective in the treatment of septic shock. The efficiency of two types of mAbs including E5 (XoMA, Berkeley, CA), a murine IgM mAb, and HA-1A (Centoxin), a human IgM, have been shown in patients with sepsis [81, 82].

Using therapeutic mAbs is also shown in the treatment of viral infections. Cytomegalovirus (CMV) could affect immunocompromised individuals, including patients with AIDS and those undergoing organ transplants. CMV proteins could be targeted by mAbs [83]. A humanized mAb against gpUL75 (gH), a glycoprotein of CMV, could interact with several strains of virus and may be considered as an appropriate agent for the treatment of patients with CMV infection [84].

A mAb named Palivizumab has been approved for Respiratory syncytial virus (RSV) infection which causes severe lower respiratory tract disorder [85]. Using mAbs has also been reported for the treatment of HSV infections.

Currently, several types of monoclonal antibodies have been designed for the treatment of patients with coronavirus disease-2019 (COVID-19) [86]. Bamlanivimab (LY-CoV555 or LY3819253) is a human IgG1 mAb against the SARS-CoV-2 spike (S) protein and could block viral entry into human cells [87]. Despite the authorization by FDA for emergency use for patients with positive SARS-CoV-2 viral test, Bamlanivimab has not been approved yet.

4.6 Monoclonal antibodies in the diagnostic assays

High specificity and high affinity binding properties of monoclonal antibodies make them effective biological reagents in immunodiagnostic assays. They can be used for diagnosis of infectious diseases and detection of certain antigens or in serological assessments for detection of antibodies against a certain antigen [88]. Monoclonal antibodies are widely used in several immunodiagnostic assays including immunohistochemistry (IHC) or immunocytochemistry (ICC), Enzyme-linked Immunosorbent Assay (ELISA), western blot, immunodot blot, radio immuno assay (RIA), Immunofluorescence (IF), flow cytometry, and microscopy (electron, fluorescence, confocal) [89].

In all methods, detection of the specific antigens on the tissue sections, cell surface, or in the homogenized sample needs the interaction between specific mAbs and the target antigen. To visualize this interaction, either the primary antibody or secondary antibody must be labeled. Totally, the primary antibody is labeled in the direct methods (such as direct ELISA, IF, and RIA) in which the antibody directly interacts with antigens immobilized on a solid tissue or on a surface [89–91]. In the indirect methods, two types of antibodies have been used. The primary antibodies are fixed on a surface and could capture antigen of interest, and then secondary antibody could interact with this complex. In these methods, the secondary antibody is labeled, allowing for signal detection. Various labels could be used, such as fluorescent molecules, enzymes, or radioisotopes. Fluorescent labeling requires a fluorescence microscope, while using enzymes such as horseradish peroxidase or alkaline phosphatase results in producing a colored product after incubation with a chromogenic substrate such as diaminobenzidine (DAB) [92–94].

5. Conclusion

The stability and flexibility of antibodies and their effector functions are important factors that determine the applicability of immunoglobulins for the development of therapeutic mAbs. The majority of the clinically available mAbs are IgG. High specificity and high affinity binding properties of monoclonal antibodies make them useful biological drugs for the treatment of a variety of disorders including autoimmunity, malignancies, and asthma. They can be used to diagnose infectious disorders and identify specific antigens, as well as in serological tests to detect antibodies against specific antigens. In addition, monoclonal antibodies are widely used in several immunodiagnostic assays with high sensitivity and specificity. Consequently, due to their important functions in both diagnosis and treatment of diseases, monoclonal antibodies have become popular molecules, particularly in medicine.

Conflict of interest

The authors declare no conflict of interest.

Author details

Mona Sadeghalvad[1,2,3] and Nima Rezaei[1,2,3]*

1 Department of Immunology, School of Medicine, Tehran University of Medical Sciences, Tehran, Iran

2 Research Center for Immunodeficiencies, Children's Medical Center, Tehran University of Medical Sciences, Tehran, Iran

3 Network of Immunity in Infection, Malignancy and Autoimmunity (NIIMA), Universal Scientific Education and Research Network (USERN), Tehran, Iran

*Address all correspondence to: rezaei_nima@tums.ac.ir

References

[1] Forthal DN. Functions of Antibodies. Microbiol Spectr. 2014;2(4):AID-0019-2014. doi: 10.1128/microbiolspec. properties

[2] Liu JK. The history of monoclonal antibody development - Progress, remaining challenges and future innovations. Ann Med Surg (Lond). 2014;3(4):113-116. doi:10.1016/j.amsu. 2014;09:001

[3] Lu RM, Hwang YC, Liu IJ, Lee CC, Tsai HZ, Li HJ, Wu HC. Development of therapeutic antibodies for the treatment of diseases. J Biomed Sci. 2020;27(1):1. doi: 10.1186/s12929-019-0592-z.

[4] Chiu ML, Goulet DR, Teplyakov A, Gilliland GL. Antibody Structure and Function: The Basis for Engineering Therapeutics. Antibodies (Basel). 2019;8(4):55. doi:10.3390/antib8040055

[5] Singh S, Kumar NK, Dwiwedi P, Charan J, Kaur R, Sidhu P, Chugh VK. Monoclonal Antibodies: A Review. Curr Clin Pharmacol. 2018;13(2):85-99. doi: 10.2174/1574884712666170809124728.

[6] Irani V, Guy AJ, Andrew D, Beeson JG, Ramsland PA, Richards JS. Molecular properties of human IgG subclasses and their implications for designing therapeutic monoclonal antibodies against infectious diseases. Mol Immunol. 2015 Oct;67(2 Pt A):171-82. doi: 10.1016/j.molimm.2015.03.255.

[7] Ryman JT, Meibohm B. Pharmacokinetics of Monoclonal Antibodies. CPT Pharmacometrics Syst Pharmacol. 2017;6(9):576-588. doi:10.1002/psp4.12224.

[8] Parray HA, Shukla S, Samal S, et al. Hybridoma technology a versatile method for isolation of monoclonal antibodies, its applicability across species, limitations, advancement and future perspectives. Int Immunopharmacol. 2020;85:106639. doi:10.1016/j.intimp.2020.106639

[9] Liu YD, Wang T, Chou R, Chen L, Kannan G, Stevenson R, Goetze AM, Jiang XG, Huang G, Dillon TM, Flynn GC. IgG2 disulfide isoform conversion kinetics. Mol Immunol. 2013 Jun;54(2):217-226. doi: 10.1016/j. molimm.2012.12.005.

[10] Holzlöhner P, Hanack K. Generation of Murine Monoclonal Antibodies by Hybridoma Technology. J Vis Exp. 2017 Jan 2;(119):54832. doi: 10.3791/54832.

[11] Kumar R, Parray HA, Shrivastava T, Sinha S, Luthra K. Phage display antibody libraries: A robust approach for generation of recombinant human monoclonal antibodies. Int J Biol Macromol. 2019 Aug 15;135:907-918. doi: 10.1016/j.ijbiomac.2019.06.006.

[12] Deantonio C, Cotella D, Macor P, Santoro C, Sblattero D. Phage display technology for human monoclonal antibodies. Methods Mol Biol. 2014;1060:277-295. doi: 10.1007/978-1-62703-586-6_14.

[13] Yousefi H, Yuan J, Keshavarz-Fathi M, Murphy JF, Rezaei N. Immunotherapy of cancers comes of age. Expert Rev Clin Immunol. 2017;13(10):1001-1015. doi: 10.1080/1744666X.2017.1366315.

[14] Foon KA, Yannelli J, Bhattacharya-Chatterjee M. Colorectal cancer as a model for immunotherapy. Clin Cancer Res. 1999 Feb;5(2):225-236. PMID: 10037169.

[15] Rezahosseini O, Hanaei S, Hamadani M, Keshavarz-Fathi M, Rezaei N. The promising role of monoclonal antibodies for immunotherapy of the HIV-associated cancer, non-Hodgkin lymphoma. Int

Rev Immunol. 2018;37(3):165-173. doi: 10.1080/08830185.2017.1405396.

[16] Baldo B.A. (2016) Monoclonal Antibodies Approved for Cancer Therapy. In: Safety of Biologics Therapy. Springer, Cham. doi.org/10.1007/978-3-319-30472-4_3

[17] Takei J, Kaneko MK, Ohishi T, Kawada M, Harada H, Kato Y. A novel anti-EGFR monoclonal antibody (EMab-17) exerts antitumor activity against oral squamous cell carcinomas via antibody-dependent cellular cytotoxicity and complement-dependent cytotoxicity. Oncol Lett. 2020;19(4):2809-2816. doi: 10.3892/ol.2020.11384.

[18] Martinelli E, De Palma R, Orditura M, De Vita F, Ciardiello F. Anti-epidermal growth factor receptor monoclonal antibodies in cancer therapy. Clin Exp Immunol. 2009;158(1):1-9. doi:10.1111/j.1365-2249.2009.03992.x.

[19] Seyedmirzaei H, Keshavarz-Fathi M, Razi S, Gity M, Rezaei N. Recent progress in immunotherapy of breast cancer targeting the human epidermal growth factor receptor 2 (HER2). J Oncol Pharm Pract. 2021;1078155221991636. doi: 10.1177/1078155221991636. Epub ahead of print. PMID: 33530866.

[20] Afshari F, Soleyman-Jahi S, Keshavarz-Fathi M, Roviello G, Rezaei N. The promising role of monoclonal antibodies for gastric cancer treatment. Immunotherapy. 201;11(4):347-364. doi: 10.2217/imt-2018-0093. PMID: 30678552.

[21] He X, Xu C. Immune checkpoint signaling and cancer immunotherapy. Cell Res. 2020;30(8):660-669. doi: 10.1038/s41422-020-0343-4.

[22] Kohler H, Pashov A, Kieber-Emmons T. The Promise of Anti-idiotype Revisited. Front Immunol. 2019;10:808. doi:10.3389/fimmu.2019.00808

[23] Wagner U., Kohler S., Reinartz S. Immunological consolidation of ovarian carcinoma recurrences with monoclonal anti-idiotype antibody ACA125: Immune responses and survival in palliative treatment. Clin Cancer Res. 2001;7:1154-1162.

[24] Foon K.A., Lutzky J., Baral R.N. Clinical and immune responses in advanced melanoma patients immunized with an anti-idiotype antibody mimicking disialoganglioside GD2. J Clin Oncol. 2000;2:376-384.

[25] Leshem Y, Pastan I. Pseudomonas Exotoxin Immunotoxins and Anti-Tumor Immunity: From Observations at the Patient's Bedside to Evaluation in Preclinical Models. Toxins (Basel). 2019;11(1):20. doi: 10.3390/toxins11010020.

[26] Sachpekidis C, Jackson DB, Soldatos TG. Radioimmunotherapy in Non-Hodgkin's Lymphoma: Retrospective Adverse Event Profiling of Zevalin and Bexxar. Pharmaceuticals (Basel). 2019;12(4):141. 20. doi:10.3390/ph12040141

[27] Siccardi AG, Buraggi GL, Callegaro L, Mariani G, Natali PG, Abbati A, Bestagno M, Caputo V, Mansi L, Masi R, et al. Multicenter study of immunoscintigraphy with radiolabeled monoclonal antibodies in patients with melanoma. Cancer Res. 1986;46(9):4817-4822. PMID: 3731128.

[28] Cruz E, Kayser V. Monoclonal antibody therapy of solid tumors: clinical limitations and novel strategies to enhance treatment efficacy. Biologics. 2019;13:33-51. doi:10.2147/BTT.S166310.

[29] Tsumoto K, Isozaki Y, Yagami H, Tomita M. Future perspectives of therapeutic monoclonal antibodies.

Immunotherapy. 2019;11(2):119-127. doi: 10.2217/imt-2018-0130.

[30] Jarosz-Griffiths HH, Holbrook J, Lara-Reyna S, McDermott MF. TNF receptor signalling in autoinflammatory diseases. Int Immunol. 2019;31(10):639-648. doi: 10.1093/intimm/dxz024.

[31] Melsheimer R, Geldhof A, Apaolaza I, Schaible T. Remicade® (infliximab): 20 years of contributions to science and medicine. Biologics. 2019;13:139-178. doi: 10.2147/BTT.S207246.

[32] Kany S, Vollrath JT, Relja B. Cytokines in Inflammatory Disease. Int J Mol Sci. 2019;20(23):6008. doi:10.3390/ijms20236008

[33] Atzeni F, Doria A, Carrabba M, Turiel M, Sarzi-Puttini P. Potential target of infliximab in autoimmune and inflammatory diseases. Autoimmun Rev. 2007;6(8):529-536. doi: 10.1016/j.autrev.2007.03.009.

[34] van den Brande J, Hommes DW, Peppelenbosch MP. Infliximab induced T lymphocyte apoptosis in Crohn's disease. J Rheumatol Suppl. 2005;74:26-30. PMID: 15742461.

[35] Senolt L. Emerging therapies in rheumatoid arthritis: focus on monoclonal antibodies. F1000Res. 2019;8:F1000 Faculty Rev-1549. doi:10.12688/f1000research.18688.1

[36] Voulgari PV, Drosos AA. Adalimumab for rheumatoid arthritis. Expert Opin Biol Ther. 2006;6(12):1349-1360. doi: 10.1517/14712598.6.12.1349.

[37] Lapadula G, Marchesoni A, Armuzzi A, Blandizzi C, Caporali R, Chimenti S, Cimaz R, Cimino L, Gionchetti P, Girolomoni G, Lionetti P, Marcellusi A, Mennini FS, Salvarani C. Adalimumab in the treatment of immune-mediated diseases. Int J Immunopathol Pharmacol. 2014;27 (1 Suppl):33-48. doi: 10.1177/03946320140270S103.

[38] Mazumdar S, Greenwald D. Golimumab. MAbs. 2009;1(5):422-431. doi:10.4161/mabs.1.5.9286.

[39] Chimenti MS, Saraceno R, Chiricozzi A, Giunta A, Chimenti S, Perricone R. Profile of certolizumab and its potential in the treatment of psoriatic arthritis. Drug Des Devel Ther. 2013;7:339-348. doi:10.2147/DDDT.S31658.

[40] Migliorini P, Italiani P, Pratesi F, Puxeddu I, Boraschi D. The IL-1 family cytokines and receptors in autoimmune diseases. Autoimmun Rev. 2020;19(9):102617. doi: 10.1016/j.autrev.2020.102617.

[41] Dinarello CA, Simon A, van der Meer JW. Treating inflammation by blocking interleukin-1 in a broad spectrum of diseases. Nat Rev Drug Discov. 2012;11(8):633-652. doi:10.1038/nrd3800

[42] Dhimolea E. Canakinumab. MAbs. 2010;2(1):3-13. doi:10.4161/mabs.2.1.10328.

[43] Lamb YN, Deeks ED. Sarilumab: A Review in Moderate to Severe Rheumatoid Arthritis. Drugs. 2018;78(9):929-940. doi: 10.1007/s40265-018-0929-z.

[44] Jung JY, Kim MY, Suh CH, Kim HA. Off-label use of tocilizumab to treat non-juvenile idiopathic arthritis in pediatric rheumatic patients: a literature review. Pediatr Rheumatol Online J. 2018;16(1):79. doi: 10.1186/s12969-018-0296-z.

[45] Perosa F, Favoino E, Caragnano MA, Prete M, Dammacco F. CD20: a target antigen for immunotherapy of autoimmune diseases. Autoimmun Rev. 2005;4(8):526-531. doi: 10.1016/j.autrev.2005.04.004.

[46] Gómez VJ, Carrión-Barberá I, Salman Monte TC, Acosta A, Torrente-Segarra V, Monfort J. Effectiveness and Safety of Rituximab in Systemic Lupus Erythematosus: A Case Series Describing the Experience of 2 Centers. Reumatol Clin. 2020;16(5 Pt 2):391-395. doi: 10.1016/j.reuma.2018.08.012.

[47] Vargas Lebrón C, Ruiz Montesino MD, Moreira Navarrete V, Toyos Sainz de Miera FJ. Treatment With Rituximab in Juvenile Dermatomyositis: Effect on Calcinosis. Reumatol Clin. 2020;16(5 Pt 1):368-370. doi: 10.1016/j.reuma.2018.06.010.

[48] Murakhovskaya I. Rituximab Use in Warm and Cold Autoimmune Hemolytic Anemia. J Clin Med. 2020;9(12):4034. doi:10.3390/jcm9124034.

[49] Wakim M, Shah A, Arndt PA, Garratty G, Weinberg K, Hofstra T, Church J. Successful anti-CD20 monoclonal antibody treatment of severe autoimmune hemolytic anemia due to warm reactive IgM autoantibody in a child with common variable immunodeficiency. Am J Hematol. 2004;76(2):152-155. doi: 10.1002/ajh.20072.

[50] Abe K, Ishikawa Y, Ishikawa J, Fujiwara M, Kita Y. Successful treatment of a patient with refractory immune thrombocytopenic purpura in systemic lupus erythematosus with rituximab. Immunol Med. 2019;42(4):185-188. doi: 10.1080/25785826.2019.1696644.

[51] Tavakolpour S, Alesaeidi S. Rituximab as an effective and probably safe treatment for granulomatosis with polyangiitis (Wegener's Granulomatosis). Int Immunopharmacol. 2019;70:67-68. doi: 10.1016/j.intimp.2019.01.029.

[52] Castillo J, Milani C, Mendez-Allwood D. Ofatumumab, a second-generation anti-CD20 monoclonal antibody, for the treatment of lymphoproliferative and autoimmune disorders. Expert Opin Investig Drugs. 2009;18(4):491-500. doi: 10.1517/13543780902832679.

[53] Voge NV, Alvarez E. Monoclonal Antibodies in Multiple Sclerosis: Present and Future. Biomedicines. 2019;7(1):20. doi:10.3390/biomedicines7010020.

[54] Fox E, Lovett-Racke AE, Gormley M, Liu Y, Petracca M, Cocozza S, Shubin R, Wray S, Weiss MS, Bosco JA, Power SA, Mok K, Inglese M. A phase 2 multicenter study of ublituximab, a novel glycoengineered anti-CD20 monoclonal antibody, in patients with relapsing forms of multiple sclerosis. Mult Scler. 202;27(3):420-429. doi: 10.1177/1352458520918375.

[55] Abdo AIK, Tye GJ. Interleukin 23 and autoimmune diseases: current and possible future therapies. Inflamm Res. 2020;69(5):463-480. doi: 10.1007/s00011-020-01339-9.

[56] Kuwabara T, Ishikawa F, Kondo M, Kakiuchi T. The Role of IL-17 and Related Cytokines in Inflammatory Autoimmune Diseases. Mediators Inflamm. 2017;2017:3908061. doi:10.1155/2017/3908061.

[57] Harjunpää H, Llort Asens M, Guenther C, Fagerholm SC. Cell Adhesion Molecules and Their Roles and Regulation in the Immune and Tumor Microenvironment. Front Immunol. 2019;10:1078. doi:10.3389/fimmu.2019.01078

[58] Zhao Y, Su H, Shen X, Du J, Zhang X, Zhao Y. The immunological function of CD52 and its targeting in organ transplantation. Inflamm Res. 2017;66(7):571-578. doi: 10.1007/s00011-017-1032-8.

[59] Vanhove B. Anticorps monoclonaux en transplantation [Monoclonal

antibodies in organ transplantation]. Med Sci (Paris). 2009;25(12):1121-1125. doi: 10.1051/medsci/200925121121.

[60] Ramachandran V, Kolli SS, Strowd LC. Review of Graft-Versus-Host Disease. Dermatol Clin. 2019;37(4):569-582. doi: 10.1016/j.det.2019.05.014.

[61] Ji SQ, Chen HR, Yan HM, Wang HX, Liu J, Zhu PY, Xiao MH, Xun CQ. Anti-CD25 monoclonal antibody (basiliximab) for prevention of graft-versus-host disease after haploidentical bone marrow transplantation for hematological malignancies. Bone Marrow Transplant. 2005;36(4):349-354. doi: 10.1038/sj.bmt.1705046.

[62] Blaes AH, Peterson BA, Bartlett N, Dunn DL, Morrison VA. Rituximab therapy is effective for posttransplant lymphoproliferative disorders after solid organ transplantation: results of a phase II trial. Cancer. 2005;104(8):1661-1667. doi: 10.1002/cncr.21391.

[63] Martin X, Da Silva M, Virieux SR, Hadj Aissa A, Buffet R, Tiollier J, Dubernard JM. Protective effect of an anti-LFA 1 monoclonal antibody (odulimomab) on renal damage due to ischemia and kidney autotransplantation. Transplant Proc. 2000;32(2):481. doi: 10.1016/s0041-1345(00)00849-6.

[64] Carswell CI, Plosker GL, Wagstaff AJ. Daclizumab: a review of its use in the management of organ transplantation. BioDrugs. 2001;15(11):745-773. doi: 10.2165/00063030-200115110-00005.

[65] Matucci A, Vultaggio A, Maggi E, Kasujee I. Is IgE or eosinophils the key player in allergic asthma pathogenesis? Are we asking the right question? Respir Res. 2018;19(1):113. doi: 10.1186/s12931-018-0813-0.

[66] Hu J, Chen J, Ye L, Cai Z, Sun J, Ji K. Anti-IgE therapy for IgE-mediated allergic diseases: from neutralizing IgE antibodies to eliminating IgE+ B cells. Clin Transl Allergy. 2018;8:27. doi:10.1186/s13601-018-0213-z.

[67] Loureiro CC, Amaral L, Ferreira JA, et al. Omalizumab for Severe Asthma: Beyond Allergic Asthma. Biomed Res Int. 2018;2018:3254094. doi:10.1155/2018/3254094.

[68] Khorasanizadeh M, Eskian M, Assa'ad AH, Camargo CA Jr, Rezaei N. Efficacy and Safety of Benralizumab, a Monoclonal Antibody against IL-5Rα, in Uncontrolled Eosinophilic Asthma. Int Rev Immunol. 2016;35(4):294-311. doi: 10.3109/08830185.2015.1128901.

[69] Thibodeaux Q, Smith MP, Ly K, Beck K, Liao W, Bhutani T. A review of dupilumab in the treatment of atopic diseases. Hum Vaccin Immunother. 2019;15(9):2129-2139. doi: 10.1080/21645515.2019.1582403.

[70] Pelaia C, Calabrese C, Vatrella A, Busceti MT, Garofalo E, Lombardo N, Terracciano R, Pelaia G. Benralizumab: From the Basic Mechanism of Action to the Potential Use in the Biological Therapy of Severe Eosinophilic Asthma. Biomed Res Int. 2018 May 10;2018:4839230. doi: 10.1155/2018/4839230.

[71] Marone G, Granata F, Pucino V, Pecoraro A, Heffler E, Loffredo S, Scadding GW, Varricchi G. The Intriguing Role of Interleukin 13 in the Pathophysiology of Asthma. Front Pharmacol. 2019;10:1387. doi: 10.3389/fphar.2019.01387.

[72] Li H, Wang K, Huang H, Cheng W, Liu X. A meta-analysis of anti-interleukin-13 monoclonal antibodies for uncontrolled asthma. PLoS One. 2019;14(1):e0211790. doi:10.1371/journal.pone.0211790.

[73] Hanania NA, Noonan M, Corren J, Korenblat P, Zheng Y, Fischer SK,

Cheu M, Putnam WS, Murray E, Scheerens H, Holweg CT, Maciuca R, Gray S, Doyle R, McClintock D, Olsson J, Matthews JG, Yen K. Lebrikizumab in moderate-to-severe asthma: pooled data from two randomised placebo-controlled studies. Thorax. 2015;70(8):748-756. doi: 10.1136/thoraxjnl-2014-206719.

[74] Hanania NA, Korenblat P, Chapman KR, Bateman ED, Kopecky P, Paggiaro P, Yokoyama A, Olsson J, Gray S, Holweg CT, Eisner M, Asare C, Fischer SK, Peng K, Putnam WS, Matthews JG. Efficacy and safety of lebrikizumab in patients with uncontrolled asthma (LAVOLTA I and LAVOLTA II): replicate, phase 3, randomised, double-blind, placebo-controlled trials. Lancet Respir Med. 2016;4(10):781-796. doi: 10.1016/S2213-2600(16)30265-X.

[75] Noonan M, Korenblat P, Mosesova S, Scheerens H, Arron JR, Zheng Y, Putnam WS, Parsey MV, Bohen SP, Matthews JG. Dose-ranging study of lebrikizumab in asthmatic patients not receiving inhaled steroids. J Allergy Clin Immunol. 2013;132(3):567-574.e12. doi: 10.1016/j.jaci.2013.03.051.

[76] Popovic B, Breed J, Rees DG, Gardener MJ, Vinall LM, Kemp B, Spooner J, Keen J, Minter R, Uddin F, Colice G, Wilkinson T, Vaughan T, May RD. Structural Characterisation Reveals Mechanism of IL-13-Neutralising Monoclonal Antibody Tralokinumab as Inhibition of Binding to IL-13Rα1 and IL-13Rα2. J Mol Biol. 2017;429(2):208-219. doi: 10.1016/j.jmb.2016.12.005.

[77] Oh CK, Leigh R, McLaurin KK, Kim K, Hultquist M, Molfino NA. A randomized, controlled trial to evaluate the effect of an anti-interleukin-9 monoclonal antibody in adults with uncontrolled asthma. Respir Res. 2013;14(1):93. doi: 10.1186/1465-9921-14-93.

[78] Corren J, Parnes JR, Wang L, Mo M, Roseti SL, Griffiths JM, van der Merwe R. Tezepelumab in Adults with Uncontrolled Asthma. N Engl J Med. 2017;377(10):936-946. doi: 10.1056/NEJMoa1704064.

[79] Menzies-Gow A, Wechsler ME, Brightling CE. Unmet need in severe, uncontrolled asthma: can anti-TSLP therapy with tezepelumab provide a valuable new treatment option?. Respir Res. 2020;21(1):268. doi:10.1186/s12931-020-01505-x.

[80] Thompson K, Venkatesh B, Finfer S. Sepsis and septic shock: current approaches to management. Intern Med J. 2019;49(2):160-170. doi: 10.1111/imj.14199.

[81] Minasyan H. Sepsis: mechanisms of bacterial injury to the patient. Scand J Trauma Resusc Emerg Med. 2019;27(1):19. doi:10.1186/s13049-019-0596-4.

[82] Berger M, Shankar V, Vafai A. Therapeutic applications of monoclonal antibodies. Am J Med Sci. 2002;324(1):14-30. doi:10.1097/00000441-200207000-00004

[83] Ohlin M, Söderberg-Nauclér C. Human antibody technology and the development of antibodies against cytomegalovirus. Mol Immunol. 2015;67(2 Pt A):153-70. doi: 10.1016/j.molimm.2015.02.026.

[84] Hamilton AA, Manuel DM, Grundy JE, Turner AJ, King SI, Adair JR, White P, Carr FJ, Harris WJ. A humanized antibody against human cytomegalovirus (CMV) gpUL75 (gH) for prophylaxis or treatment of CMV infections. J Infect Dis. 1997;176(1):59-68. doi: 10.1086/514040.

[85] Groves HE, Jenkins L, Macfarlane M, Reid A, Lynn F, Shields MD. Efficacy and long-term outcomes of palivizumab prophylaxis to

prevent respiratory syncytial virus infection in infants with cystic fibrosis in Northern Ireland. Pediatr Pulmonol. 2016;51(4):379-385. doi: 10.1002/ppul.23376.

[86] Jahanshahlu L, Rezaei N. Monoclonal antibody as a potential anti-COVID-19. Biomed Pharmacother. 2020;129:110337. doi: 10.1016/j.biopha.2020.110337.

[87] Jones BE, Brown-Augsburger PL, Corbett KS, Westendorf K, Davies J, et al. LY-CoV555, a rapidly isolated potent neutralizing antibody, provides protection in a non-human primate model of SARS-CoV-2 infection. bioRxiv [Preprint]. 2020 Oct 1. doi: 10.1101/2020.09.30.318972.

[88] Wilson K & Walker J. Principles and Techniques of Biochemistry and Molecular Biology. Director; 2010. 802p.

[89] Safi MA. An overview of various labeled assays used in medical laboratory diagnosis. Immune and non-immune assays. Saudi Med J. 2010;31(4):359-368. PMID: 20383411.

[90] Aoki V, Sousa JX Jr, Fukumori LM, Périgo AM, Freitas EL, Oliveira ZN. Direct and indirect immuno fluorescence. An Bras Dermatol. 2010;85(4):490-500. doi: 10.1590/s0365-05962010000400010.

[91] Shrivastav TG, Kanaujia PK. Direct radioimmunoassay for the measurement of serum testosterone using 3H as label. J Immunoassay Immunochem. 2007;28(2):127-136. doi: 10.1080/15321810701211700.

[92] Taylor CR. Immunohistochemistry in surgical pathology: Principles and practice In: Day CE (ed) Histopathology: methods and protocols. Methods in molecular biology. Springer; 2014. pp 81-110.

[93] Lin AV. Indirect ELISA. Methods Mol Biol. 2015;1318:51-59. doi: 10.1007/978-1-4939-2742-5_5.

[94] Silverstein AM. Labeled antigens and antibodies: the evolution of magic markers and magic bullets. Nat Immunol. 2004;5(12):1211-1217. doi: 10.1038/ni1140.

Chapter 2

Alternative Methods to Animal Use for Monoclonal Antibody Generation and Production

Jane Zveiter de Moraes, Barbara Hamaguchi, Camila Braggion, Enzo Speciale, Fernanda Cesar, Gabriela Soares, Juliana Osaki, Rodrigo Aguiar and Tauane Pereira

Abstract

Monoclonal antibody (mAb) has broad applicability in research, diagnosis, and treatment. After the introduction of hybridoma technology in 1975, the mAb market has increased dramatically, moving a large industry of more than US$ 140 billions in 2020. In 1954, the concept of the 3R's was proposed and much changed the animal use scenario, including the recent ban on inducing ascites in mice for the production of mAb. In light of this, the generation and production of antibodies had to be reassessed. In this chapter, we present an overview of the main alternative technologies to the use of animals in the generation and production of mAb. Antibody display libraries and *in silico* modeling are very promising technologies that may provide mAb genetic constructs that, in the sequence, may be expressed on mammalian, bacterial, yeast or plant systems. Although the total replacement of the use of animals in the entire process is not currently feasible, it is possible to find ways to reduce and refine the use of animals in obtaining and producing mAb.

Keywords: monoclonal antibody, alternative methods, antibody generation, antibody display libraries, *in silico* antibody modeling, antibody expression systems

1. Introduction

Animals have been used for research applications since the early centuries after Christ [1]. This practice has always been controversial. However, only in 1870, after discovering that animals feel pain, the theme began to be reconsidered [2]. Yet, the first significant milestone involving ethical issues in using animals for research occurred in 1954, when Charles Hume and William Russel proposed the concept of the 3Rs. They advocated "Replacing, Reducing and Refining" the use of animals to minimize pain or stress whenever possible [3]. This conception was further extended to the 6Rs to include "Read across", referring to the critical analysis of new results, "Relevance", which concerns ethical and educational visions, including good laboratory practices, and finally "Roadmaps," which evolves planning, communication, conference and technical implementation policies [4]. Recently, the American

Anti-Vivisection Society (AAVS) banned the production of ascites in animals, launching the "antibodies without animals" campaign [5, 6].

In view of this, the conventionally used strategies to generate and produce monoclonal antibodies (mAbs), initially proposed by Köhler and Milstein in 1975 [7] as the hybridoma technology, had to be reconsidered. This methodology is based on the immunization of animals with the antigen of interest, followed by the fusion of B lymphocytes with myeloma cells, resulting in the formation of hybridomas. After cloning and selection, the antibody-secreting stable monoclonal cell lines were used to produce mAb-enriched ascitic liquid [8], a now-banned practice. For many years, mAbs, which have wide applicability in research, diagnosis, and treatment, were generated and produced with this methodology. In this chapter, the main technologies that emerged as alternatives to the use of animals for the generation and production of mAbs are discussed.

1.1 Overview of antibody structure and most common formats

mAbs are immunoglobulin molecules with a molecular weight of ~150 kDa, made up of four polypeptide chains: one pair of identical light chains and another pair of identical heavy chains joined by disulfide and non-covalent bonds. Each chain contains a variable domain (VL and VH) at the N-terminal portion and one or three constant domains at the polypeptide's C-terminal portion. The antigen-binding fragment (Fab) has ~50 kDa and is composed of variable and constant regions of heavy and light chains. The variable fragment (Fv) includes only the two variable domains of both chains (**Figure 1a**) [9]. Each variable region is composed

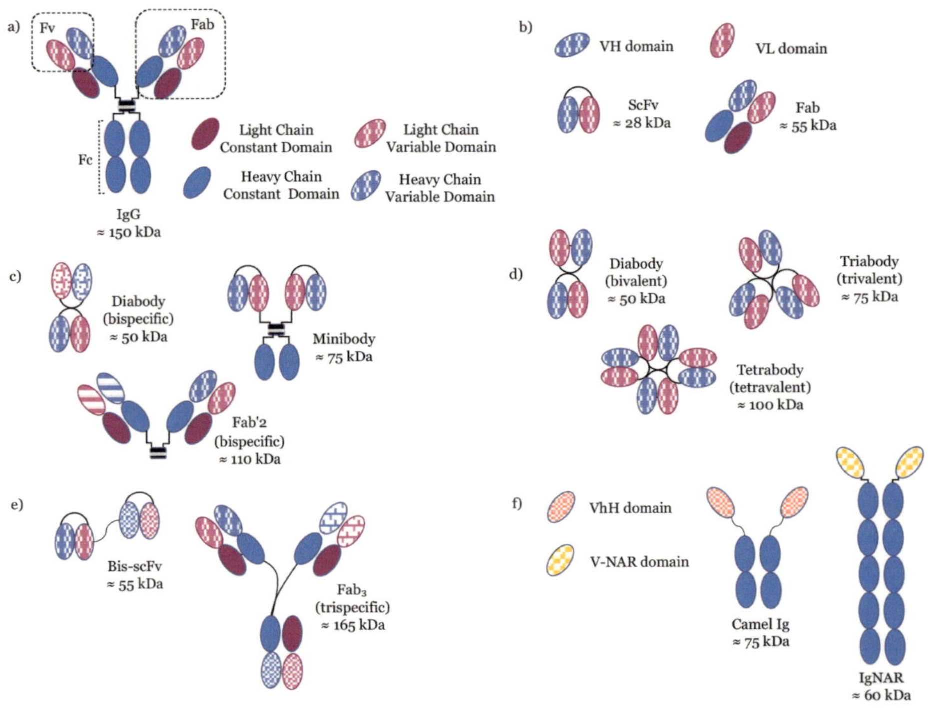

Figure 1.
Schematic representation of different antibody formats: (a) the classic IgG and its respective regions and chains; (b) monovalent antigen-binding formats: Fab and scFv; (c) bivalent antigen-binding formats: Fab'2, diabody and minibody. Different textures indicate different antigen specificities; (d) structures composed of scFv can form diabody, triabody and tetrabody; (e) other possible formats that can be constructed; (f) camelid Ig and shark IgNAR molecules.

of six complementarity determining regions (CDRs), also known as hypervariable regions: three in the light chain (L1, L2, and L3) and three in the heavy chain (H1, H2, and H3), CDR H3 being the most variable in length, sequence, and structure. These regions promote the high specificity of functional binding of mAbs with the target antigen [10]. In spite of hypervariable regions, it should be noted that these regions assume conformations in the form of loops because of the presence of some conserved residues [11].

There are several formats of antibodies used for therapeutic, diagnostic, and basic research purposes. Smaller formats were initially generated by the removal of the constant domain (Fc) by proteolysis, with the enzymes papain and pepsin, and later by genetic manipulations, giving rise to monovalent antigen-binding formats: Fab and scFv (variable single-chain fragment) (**Figure 1b**) or bivalent formats such as Fab'2, diabody, minibody, among others (**Figure 1c**). These antibodies present characteristics in comparison to complete mAbs that may be useful depending on the application: besides retaining the antigen-binding affinity of the parental antibody, they have a reduced serum half-life and are less immunogenic [12, 13]. In addition to conventional shapes, camelids and sharks produce unusual antibodies composed only of heavy chains, with just a single domain in its variable antigen-binding site. In Camelidae, it is called variable domain of heavy chain antibodies (VhHs), while in some cartilaginous fish, like sharks, it is called new variable antigen receptor (V-NARs). The smaller sizes of VhHs and V-NARs allow them to be good candidates as biotechnological tools (**Figure 1f**) [14–16].

Fab favors the crystallization of several proteins facilitating the determination of their three-dimensional structure [15]. Therapeutic Fabs have been available since 1994, starting with the chimeric abciximab Fab (ReoPro), used as an antiplatelet agent. Ranibizumab (Lucentis) was approved by Food and Drug Administration (FDA) in 2006 for treatment of age-related macular degeneration, certolizumab pegol (Cimzia) was approved in 2009 for rheumatoid arthritis, and several other Fabs are now in clinical and pre-clinical trials. This format is also useful for diagnostic imaging, like arcitumomab (CEA-scan) approved in 1996 for colorectal cancer screening [12, 17, 18].

Another variant is scFv antibodies, with a molecular mass of ~30 kDa, composed of VH and VL domains joined with a peptide ligand. These structures can be presented as dimers, trimers, tetramers (**Figure 1d**) or other formats (**Figure 1e**) through genetic or chemical manipulations. Diabodies can present two identical antigen-binding sites, when it is called a bivalent diabody, or have two different antigen-binding sites, a bispecific diabody [19, 20]. Numerous scFvs have been constructed against haptens, proteins, carbohydrates, receptors, tumor antigens, and viruses for applications in therapy and diagnosis [21]. In 2019, the FDA approved the humanized scFv brolucizumab (Beovu) to treat neovascular age-related macular degeneration. Other scFvs are already in a pre-clinical study for targeted cancer therapy [22].

2. Alternative methods for monoclonal antibody generation

2.1 Antibody display libraries

Antibody display libraries are powerful tools to isolate high-affinity antibodies for therapeutic and/or diagnostics applications. They can be divided into two groups: cell surface and cell-free display libraries. In the first case, the antibody is expressed on the surface of bacteria, yeasts, or mammalian cells, using own internal machinery [22] and, in the second, extracts of prokaryotic or eukaryotic cells, mainly from rabbit or wheat germ, are used to transcribe and translate genetic information contained in the library [23, 24]. Although they use different paths, these methodologies have a common property: they are potential alternative methods to the use of animals for antibody generation.

These methodologies involve two main steps: the construction of the library and the selection of the antibody of interest. According to the origin of the genetic material used in their construction, the libraries are classified as naïve, immune, synthetic, or semi-synthetic. The immune library is obtained from animals or humans that have been immunized and developed antibodies against a particular antigen [25, 26]. The other types are known as universal libraries: naïve is cloned from non-immunized donors; the semisynthetic is created using both naturally and synthetically (*in silico*) randomized CDRs, which increases the diversity of the library without requiring a large number of donors; and synthetic, based only in *in silico* design and gene synthesis to optimize individual amino acids, hence expanding its diversity, expression, and stability [27].

2.1.1 Cell surface display

Among the display libraries, the phage display stands out for being the first described and currently more used in the generation of antibodies [28]. The technique was developed by George Smith in 1985 and it uses the bacteriophage's ability to infect bacteria. In this way, a foreign DNA sequence is inserted into the genes III or VIII, which encode the pIII and pVIII coat proteins, respectively. The recombinant protein is displayed on the outer surface of the phage as a fusion protein in an immunologically accessible form [29]. The displayed antibodies can be of either Fab or scFv formats. The phage display library is generated by assembling DNA sequences that encode antibody fragments in the phage or phagemid vectors. The phage vector has a complete phage genome, but it is not effective for large proteins. A phagemid vector is a plasmid that contains phage coat gene (gIII or gVIII), and phage and plasmid's origin of replication. The vectors are used to transform *E. coli*. The phage vector has all the ability to produce phage particles and display the fusion antibody, while the phagemid needs to infect the bacterium with a helper phage to enable the recombinant DNA package, as a single-strand DNA into virion particles, and to display the antibody fragments [25, 28, 30]. The screening of displayed antibodies is performed by biopanning, a process in which the phages are incubated with the immobilized antigen, and the non-binding phages are removed by extensive washing. The bound phages are then eluted and enriched by reinfection of *E. coli* and thus successive rounds of selection can be carried out as many times as needed [31].

Bacterial display, an alternative to phage display, allows libraries of greater diversity. In this system, the expression of recombinant proteins is easier and the transformation by DNA is more efficient than phage display. The methodology is fast, easy to handle, and eliminates the stage of infection by the phage. The library can be displayed on the membrane in the periplasmic space or fused to the filament flagellar or fimbrial adhesin proteins [32, 33]. To generate the bacterial library, the sequences of DNA, encoding scFv or Fab fragments, are inserted into the appropriate display vector used to transform *E. coli*, the most common bacterial strain used in this technique [34]. The target antigen, adsorbed on magnetic beads or fluorophore-labeled, is used for the screening of antibody libraries by cell sorting [35].

In the cases of yeast and mammalian displays, eukaryotic systems, folding and post-translational modifications, which are relevant to the function and stability of the antibody, are more effective when compared to what occurs in the prokaryotic system [22]. Briefly, the yeast library is created by linking the antibody gene sequence into suitable yeast display vectors. The transformed yeast by the plasmid generated is induced to express the recombinant library. The system can display scFv, Fab, or full-length antibody formats [36], that are expressed in fusion with anchor proteins of the glycosylphosphatidylinositol (GPI) family, such as α-agglutinin and a-agglutinin. The screening is performed by cell sorting [37].

When using the mammalian system, the post-translational modifications are still more effective than those possible in the yeast system. The antibody library can be usually displayed on the surface of Human Embryonic Kidney 293 T (HEK 293 T) and Chinese Hamster Ovary (CHO) cells after transient or stable transformation [38]. The antibody expressed is fused to the transmembrane domain of human platelet-derived growth factor receptor (PDGFR), which anchors the antibody on the outer surface of the cell membrane [39]. The library screening is performed as already described for other cell-surface display systems [40].

2.1.2 Cell-free display

The cell-free display libraries, unlike previously described ones, do not depend on the efficiency of transduction or transfection. Among them, the ribosome and mRNA systems have been the most described, offering around 10^{12}–10^{14} variants, a wider diversity over other display techniques, like phage (around 10^{9}), eukaryotic (10^{6}–10^{7}), and prokaryotic (10^{8}–10^{10}) systems [41].

While the ribosome display connects nascent proteins to their encoding mRNA through the generation of stable protein–ribosome–mRNA complexes, the mRNA system uses an antibiotic, puromycin, that mimics the structure of an aminoacylated tRNA, to modify mRNA that also is linked to its respective nascent protein. In a brief description, a ribosome system display construct is designed to be used with cell extracts to allow the downstream mRNA synthesis. In the construct, it must be present a ribosome binding site to the start codon where protein synthesis begins, for recruitment and pairing of the ribosome. The open reading frame is followed by the library of binding proteins and a spacer. The spacer provides flexibility to the display library in order to fold outside of the ribosome tunnel. Another important point is that the ribosome is stalled at the 3′-end by deleting the stop codon to couple the nascent polypeptide with its encoding mRNA [23, 42, 43].

The mRNA display stands out in relation to the ribosome display for offering a large degree of control over experimental conditions [41]. Briefly, the DNA antibody library is *in vitro* transcribed in mRNA. In a second step, a covalent interaction between mRNA and puromycin is produced, providing after translation the formation of the mRNA-puromycin-protein complex, which is reverse transcribed into cDNA to obtain the heteroduplex cDNA-mRNA, more stable than mRNA alone. After screening, the selected cDNA is amplified by PCR. The amplified constructs are subjected to a new cycle to obtain the mRNA-puromycin-protein complexes, and then the heteroduplexes obtained are ready for a new round of screening [41, 43].

Both methodologies enable the generation of different formats of the antibodies, including the full-length ones. The selection of the antibodies of interest is performed by binding to an immobilized antigen. Ribosome and mRNA systems have gained relevance for allowing efficient and low-cost antibody production and for their advantageous ability to screen large libraries. Although promising, so far there is no commercially available antibody generated by cell-free technologies.

2.2 Antibody design via *in silico* modeling

Advances in DNA and protein sequencing techniques associated with X-ray crystallography approaches to evaluate the antibody structure at an atomic level and the increasing availability of the generated data in public domains provided a fundamental basis to the in-silico generation of mAbs.

Computer-assisted design of new mAbs consists of high-throughput algorithm analyses of antibody structures modeled from query residue sequences. These models are typically obtained by homology with precompiled antibody scaffold templates [44], which is possible because, despite the unique spatial identity of mAbs, the geometry of their variable regions is well conserved, with most CDR loops having a limited number of conformations, known as canonical classes [11]. In general, the established modeling tools coupled with refined protein–protein docking [45] and machine learning methods have been found useful for predicting the VH and VL domain arrangements and the potential antibody electrostatic complementary interface [46].

Examples of platforms available for antibody modeling are the "Prediction of ImmunoGlobulin Structure" (PIGS) [47], the Rosetta Antibody Modeling [48], and the "Web Antibody Modeling" (WAM) [49]. These servers comprise fully automated homology-based modules that predict with high accuracy the tridimensional antibody structure, including most of the hypervariable regions of the antigen-binding site [50]. An exception is the H3 loop. Unlike the other CDRs, the H3 structure has unique conformations that do not follow a canonical form and are also not found in any described protein, with ~75% of its fragments not having structural neighbors in the known non-immunoglobulin protein world [51]. Therefore, the H3 loop cannot be predicted by selecting templates from a database and this is an important obstacle for the *in silico* antibody design. Some alternative algorithms, based on candidate conformations obtained computationally and energy functions, have been developed, but they often fail to produce sub-angstrom structure models [50, 52] and the problem persists.

Other concerns also affect antibody modeling. The limited number of high-quality X-ray crystal structures of mAbs in public protein databases may not be sufficient to allow a proper antibody shape prediction [50]. Regarding the docking protocols, it should be noted that, despite the great advances in the bioinformatic field, most of the antibody algorithms still need to be optimized to consider the molecular backbone flexibility and the transient conformational changes following protein–protein interactions [53, 54]. Another relevant point is the time needed for antibody modeling. High-throughput computational design of mAbs can still be as time-consuming as experimental cellular approaches, even when well-consolidated prediction systems are used. As an example, the Rosetta Antibody server was previously found to take 570,000 CPU hours to generate ~2,000 antibody models [55].

With many challenges ahead, currently there are few reports of functional antibodies completely designed by in-silico approaches. A successful attempt in this field is the mAb described by Nimrod and co-workers, which was based on robust predictions of specific residue-residue interactions rather than modeling the entire antigen–antibody complex [56]. On the other hand, computational protocols have been used with increasing frequency to improve the physicochemical properties of previously generated mAbs, as well as to engineer humanized versions of murine full-length immunoglobulins, making them like those found in humans [57]. Molecular structure-based iterative algorithms have been shown to optimize the generation of humanized antibody scaffolds without a significant drop in affinity and specificity toward the antigen, compared to the original murine one, and with reduced occurrence of structure failures, important drawbacks commonly found following conventional humanization techniques, which are mostly guided by linear antibody residue sequences [58].

The overall computational antibody discovery scenario is promising and, although the design of new biologically active mAbs is still deeply dependent on living animals, the advances in structure prediction methods set the scene for an ongoing technological evolution that should potentially lead the future generation of these molecules using only in-silico approaches.

3. Alternative methods for mAb production

3.1 Mammalian production systems

As previously mentioned, one of the major utilities of the mammalian expression systems is to produce complex biomolecules such as antibodies that require post-translational modifications like glycosylation [38]. Though other eukaryotic systems do provide this modification, their capability of doing so is limited and might result in the addition of glycans that are not common in human proteins [59]. This event might result in misfolding and biologically inactive immunoglobulins, undesirable features in human therapeutic and diagnostic monoclonal antibodies [60]. Also, expression in prokaryotic systems might lead to contamination with endotoxins, which increases downstream processes to clear these endotoxins from the final product. Thereby, the mammalian expression systems are valuable tools to produce monoclonal antibodies as well as other proteins with proper structure and activity. Indeed, there are numerous FDA-approved mAbs produced in mammalian expression systems in contrast to prokaryotic systems and other eukaryotic cells [61].

The primary technique for mAb obtention was already originally dependent on a mammalian cell: the hybridoma cell [7]. As the high specificity of the monoclonal antibodies was making these molecules increasingly useful for various applications, a hybridoma large-scale cultivation became a great demand in the industry. Therefore, the ascites method production was no longer enough to supply bulk production, nor feasible due to ethical matters. That way, most research and diagnostic proposed mAbs are now produced *in vitro,* through the harvest and following purification of mAb-enriched media obtained in dynamic or non-dynamic cell culture systems [62].

In a therapeutic context, although the hybridoma cell lines are still responsible for the generation of more than 50% of the FDA-approved mAbs [63], these antibodies are bulk produced in other mammalian host systems [64]. This is due mainly to the highly immunogenic nature of murine mAbs for humans, demanding the antibodies to be genetically modified (humanization or generation of fragments) for human therapeutic use [65]. Besides, many of these mammalian cells had their expression machinery highly optimized for recombinant protein production [66].

In the mammalian expression system, cells are readily transfected or transduced to introduce foreign DNA that codes for the target protein and then, they are cultivated preferably in suspension in a chemically defined serum-free media [59].

The preferred mammalian cell lines for protein expression in research and industrial fields are CHO and HEK-293 cells [67]. CHO cells are dominant in heterologous protein production in industry, mainly because of advantages like the property to provide complex post-translational modifications similar to those of humans, their ease to scale-up, and for being easily adapted to grow in serum-free suspension cultures [65]. CHO cells are more suited for stable expression, for its transfection renders low yields of recombinant protein secretion in this lineage. Since establishing a stable cell line is time and labor-consuming, transient transfection is a suitable option to gather high amounts of proteins in a shorter period. In this case, HEK cells might represent an interesting option, since they are well-known for being rather suitable for transient transfection. This cell line also has rapid doubling time and grows in high-density concentrations, just like CHO cells, and presents productivity of grams of protein per liter of culture [67, 68], though they have somewhat more tendency to clump [69, 70].

The PER.C6 cells are human embryonic retinal cells, and like HEK cells are pointed out to promote human glycosylation profiles. They were projected to be grown in high-density conditions [71], with stable expression and also offer production yields similar to CHO cells, indicating that human cell lines will be more economically viable and more easily scalable options for antibody productions [72]. There is even description of a production with titers of 27 g/L of antibody, astounding yields when compared to a medium CHO cell production of around 12 g/L of antibodies [73]. Although there is still no FDA-approved mAb produced in this system, there are already some ongoing clinical and preclinical studies carried out with mAbs and other biological products purified from this system, like vaccines for influenza, HIV, and Ebola [74–77].

There are other suitable host cell lineages such as murine lymphoid cell lines like NS0 and Sp2/0-Ag14, derived from BALB/c mice plasmocytomas, corresponding for almost 25% production systems of FDA-approved monoclonal antibodies. One of their major advantages is being originated from naturally high immunoglobulin producing parental cells. Though, their murine origin is not to be underestimated, for there are reports that they do generate immunogenic glycoforms of the expressed antibodies [78].

Concerning the expression vectors for mAb production, usually the plasmids carrying the heavy and light chain genes are constructed based mainly into two kinds of systems: the dihydrofolate reductase (DHFR) system or glutamyl synthetase (GS) based system, both acting as selection markers [65]. In DHFR, selection occurs through glycine, hypoxanthine, and thymidine depletion from the cell culture medium. Selected clones are subjected to the addition of methotrexate, a folate analog that poisons the cells deficient in DHFR, obliging the cells to further synthesize the enzyme with consequent co-amplification of the IgG genes. In the GS system, the selection is done in the absence of glutamine, in a way that only cells with GS can survive by synthetizing glutamine from glutamate and ammonium. Here, the selective pressure is made through increasing doses of the GS inhibitor methionine sulfoximine, pushing the cells to amplify GS and IgG genes [79]. Promoter characteristics, inclusion of antibiotic resistance genes, transcription termination sequences [poly(A)], and translation control sequences should also be taken into account when designing these vectors [80].

In comparison to other production systems, mammalian cells are more fastidious to culture than bacteria and fungi, for they are larger and do not possess tough cell walls like other microorganisms, making them more sensitive to impurities naturally occurring from the production system itself. Having them to thrive and reproduce in culture after modifications to turn them stable and in conditions to secrete the aimed molecule with high yields is a challenge in itself [61, 67]. If one is not choosing for the transient transfection, having the stable lineages may also be costly and time-consuming.

Independent of the expression system, the correct choice of the production scale should be made accordingly to the given necessity. The simplest culture system is the static culture, consisted of T bottles with screw caps kept horizontally in an incubator. Because of its low maintenance profile and low costs, it is the most widely used culture method in the academic research context. It is possible to use this system for clone screening and determining experimental conditions, but its small-scale nature might not render enough mAb quantity for some other types of assays. An option to circumvent this matter might be the use of the rolling systems that offer a medium-scale mAb yield. In this condition, roller bottles are positioned in a rotation system that causes all cells to be in constant movement, and therefore, all of the components of the culture (cells, nutrients, dissolved gases, and metabolites) are uniformly distributed throughout the volume of the medium. This system requires gradual adaptation to cell growth in suspension, starting from very low rotation speeds [81–83].

Currently, the bulk production of mAbs in agitated bioreactors is the predominant cell culture system in the industry because it allows constant control and monitoring of the process. The area of research for innovation in these bioreactors has advanced dramatically. In general, bioreactors are used to achieve high cell densities and thus increase the production of monoclonal antibodies, biopharmaceuticals, and vaccines [84]. Different types of agitated bioreactors have been used for the cultivation of mammalian cells, both on a pilot and industrial scales.

3.2 Bacterial and yeast production systems

The use of microorganisms such as bacteria and yeasts is widely used in science for several purposes, generally related to antibiotics and probiotics [73]. However, both bacteria and yeasts have been getting space in the production of mAbs for immunological therapy due to the biopharmaceutical demand and technological advances about their ability to produce antibodies by reducing the use of animals in the manufacturing process [73]. The motivation behind investments that seek to optimize the means of production of mAbs in alternative models stems from the manufacturing disadvantages presented in the traditional method with mammalian cells, which have been predominantly employed in the expression of these antibodies due to their ability to introduce post-translational modifications similar to those human cells [85]. The mammalian expression system is expensive and time-consuming, and efforts have been made to express them in different systems. Microbial cells of yeasts and bacteria have many advantages, such as typical rapid growth, low cultivation costs, and genetics well known in the literature [86]. Microorganisms can produce high molecular weight compounds like proteins, perform highly selective reactions by their native enzymatic machinery, and also allow the repeated introduction of enzymes or immobilized cells [87]. In addition, finally, processes that use microorganisms do not generate organic and inorganic pollutants, such as mercury and toluene [88]. Still, it was complicated to produce complete antibodies in prokaryotes to the detriment of the insecurity of microbial products for human use [73]. Fortunately, the FDA published a special set of rules called "Generally Recognized as Safe" (GRAS), which guarantees the human safety of microbial products and the production of monoclonal antibodies [89]. Thus, several microorganisms were explored. In the case of gram-negative bacteria, *Escherichia coli* stands out, once it has two compartments for protein expression - the cytoplasm and the periplasmic space [86]. Gram-negative bacteria also have an oxidizing environment which allows the correct formation of disulfide bonds [90].

Whole antibodies can be produced in bacteria and this process is dependent on periplasm, which is an essential region for folding the proteins and chains that make up the structure of antibodies. Unfortunately, studies reveal very low levels of periplasm, which limits the yield for mAbs production [91]. Efforts to produce antibodies in the cytoplasm have not been successful until recently [92]. Gram-positive bacteria are more advantageous than gram-negative bacteria because they do not produce endotoxins - a highly immunogenic lipopolysaccharide (LPS) produced by gram-negative bacteria. Fewer complex eukaryotes such as yeasts have also been exploited for the production of mAbs. They have the advantage over prokaryotes in similarity with the mammalian protein expression system, allowing the expression and folding of complex proteins more easily, and yet, as well as gram-positive bacteria, do not produce endotoxins [93].

Among yeasts, *Pichia pastoris* and *Saccharomyces cerevisiae* dominate the field in the production of antibodies [86]. *S. cerevisiae* is promising due to the advantage of being well characterized, but the correct folding of chains and proteins and low yields are problems to be faced. On the other hand, *P. pastoris* does not secrete many endogenous proteins that need to be removed in the mAbs production process [94]. Yeasts have cellular glycosylation machinery, however, their proteins exhibit types of glycosylation completely different from human proteins, and this results in a significant reduction in therapeutic effector functions [95]. Whether from yeasts or bacteria, native full-length mAbs need to be glycosylated during their synthesis, but this is an obstacle that has yet to be overcome for better production efficiency in microbial hosts. The glycosylation status of the Fc region is critical for the recruitment of serum proteins from the complement system and the destruction of target cells by complement-dependent cytotoxicity (CDC) cascades [95]. This is the main reason why the method of producing mAbs in mammalian cells is still the most applied [91]. Until 2020, there are 151 recombinant therapeutic proteins approved by the FDA, one-third of them are mAbs but there are many other mAbs under development. Among these mAbs, only two are antigen-binding fragments (Fabs) that are produced in the periplasm of the bacteria *E. coli*: ranibizumab and certolizumab pegol. The first, ranibizumab, approved in 2006, is an IgG1 Fab fragment used to treat neovascular age-related macular generation and macular edema after retinal vein occlusion. Certolizumab pegol is also a humanized Fab fragment, approved in 2008 for the treatment of Crohn's disease and rheumatoid arthritis. Therefore, since the advent of mAb therapy, the biopharmaceutical industry has been investing considerable resources in new bioprocesses for the manufacture of glycosylated antibodies that attach human IgG-like glycans through alternative host expression [95].

3.3 Plant-based antibody production systems

To produce antibodies in plants, a transformation is mediated by a bacterium that infects plants, called *Agrobacterium*. The bacterium then loads the expression vector with the antibody gene, thus generating the transgenic plants that express the desired antibody. The transformed Agrobacterium is inoculated into the leaf slices of the plants. These slices regenerate in 3–4 weeks. Small shoots are then formed from the callus and transferred to a plant cultivation box *in vitro*. For the production of biomass, *in vitro* transgenic plants are transferred to a soil pot and grown in a greenhouse [96]. The most used plant systems are tobacco and alfalfa because they are the most accessible and common sources of leaf biomass. Tobacco has great advantages, such as high leaf biomass yield and rapid scaling up through easy seed production compared to other plant species. However, tobacco contains nicotine and other toxic alkaloids that need to be removed through an additional extraction step [97].

The plant system offers important advantages, such as high production capacity, low cost in the large-scale cultivation process, in addition to avoiding ethical problems associated with animals [98]. Another important advantage of using this system is found in post-translational protein modifications, which occur in plant cells in a similar way to animal cells, as well as in the correct assembly of complex molecules, such as antibodies, are aided by chaperones that mediate folding and the formation of disulfide bonds, while the addition of N-glycans is carried out by specific cellular glycosyltransferases. In fact, while core N-glycans are similar in plants and mammals, complex N-glycans show substantial differences with sialic acid [99–101].

In addition, there is a possibility to design a custom antibody glycosylation profile, and production can be enlarged simply by increasing the number of plants [102]. In comparison with the systems described earlier, the use of plants for the production of antibodies offers several irreplaceable benefits. Plants are widespread, abundant, and develop more quickly because they normally mature after a growing season. It is possible to put the product on the market quickly, which ends up decreasing the cost of production. Plants also reduce screening costs for bacterial toxins, viruses, and prions because they are less likely to introduce animal pathogens than mammalian cells or animals [98].

The disadvantages of this system are found in the low yield of protein expression, the downstream processing problems related to the extraction of proteins from leaves, and some regulatory obstacles [103].

The first pioneering study on the production of full-size IgG in plants dates back almost 30 years ago [104]. Since then, different antibody formats have been expressed in plants, such as IgA, Fab fragments, minibodies, and scFvs [103]. The first drug from plant cells to receive FDA approval for human use was the enzyme β-glucocerebrosidase, commercially called ELELYSO, indicated for the treatment of patients with a confirmed diagnosis of Type 1 Gaucher disease [105]. Thereafter, Medicago Inc. developed a quadrivalent plant-derived seasonal influenza vaccine that recently completed Phase III clinical trials [106]. A study published in September 2020 positively demonstrated the expression of a scFv 13F6 antibody with binding activity against Ebola virus-like particles in a plant system [107, 108]. Of the antibodies produced by plants, there are already 6 against viruses, 5 against tumors, and 3 against bacteria [97].

Therefore, given the data presented and the clear advantages, we can say that the plant system is quite efficient and may, in the future, be widely used in the production of antibodies both in basic research and on an industrial scale.

4. Conclusions

Bearing in mind that obtaining high specificity and affinity mAbs is not trivial, there is a great race to develop methodologies that can meet the most varied demands. An overview of the main technologies clearly shows that the total replacement of animals' use in the generation and production of mAbs is not possible for the moment. We believe that this will only be reached when the in silico technology is fully dominated. But as the implementation of alternative methods must be seen as a process, reducing and refining the use of animals are achievements. Thus, the different types of antibody display libraries represent a major breakthrough. As described, the source of genes for building the libraries may imply greater or lesser use of animals and only synthetic display libraries completely dispense the use of animals. In the same way, for the production of mAbs, several possibilities are currently available. The important thing in the production stage is that the use of ascites, a proceeding that brings pain and stress to animals, may already be eliminated in most cases. Invariably, the purpose and amount of the mAb to be produced will determine the choice of obtaining and production methodologies. Given the great utility and diversity of mAb uses, ranging from therapeutic application to essential research tools, and the wide range of technologies available today for obtaining and producing them, it seems a fact that it is always possible to choose or design a path that meets the concept of 3Rs.

Acknowledgements

The authors acknowledge the support from Conselho Nacional de Desenvolvimento Científico e Tecnológico (CNPq, Brazil), Coordenação de Aperfeiçoamento de Pessoal de Nível Superior (CAPES, Brazil), and Fundação de Amparo à Pesquisa do Estado de São Paulo (grant no. 16/14358-2; FAPESP, Brazil).

Conflict of interest

The authors declare no conflict of interest.

Author contributions

T.P. and F.C. wrote the "Overview of antibody structure and most common formats" section; J.O. and J.M. wrote the "Antibody display libraries" section; R.A. wrote the "Antibody design via in silico modeling" section; B.H., G.S., and C.B. wrote the "Mammalian production systems" section; E.S. wrote the "Bacterial and yeast production systems" section; C.B. wrote the "Plant-based antibody production systems" section; and J.M. drafted the "Conclusions" section. All authors contributed critically to the chapter preparation.

Author details

Jane Zveiter de Moraes[1*], Barbara Hamaguchi[1], Camila Braggion[1], Enzo Speciale[1], Fernanda Cesar[2], Gabriela Soares[1], Juliana Osaki[1], Rodrigo Aguiar[1] and Tauane Pereira[1]

1 Federal University of São Paulo, São Paulo, Brazil

2 São Camilo University Centre, São Paulo, Brazil

*Address all correspondence to: janezvmo@gmail.com

References

[1] Davis SL, Cheeke PR. Do domestic animals have minds and the ability to think? A provisional sample of opinions on the question. J Anim Sci. 1998 Aug;76(8):2072-2079. doi: 10.2527/1998.7682072x. PMID: 9734856.

[2] Purchase IF, Nedeva M. Editorial: Attitudes to alternatives to animal experimentation. Altern Lab Anim. 2002 May-Jun;30(3):351. doi: 10.1177/026119290203000313.

[3] Balls M. Replacement of animal procedures: alternatives in research, education and testing. Lab Anim. 1994 Jul;28(3):193-211. doi: 10.1258/002367794780681714. PMID: 7967458.

[4] SBM. Sociedade Brasileira de Metrologia from: http://metrologia.org.br/wpsite/pan-main/ [accessed: 2020-12-10]

[5] AAVS. Research Alternatives - American Anti-Vivisection Society. Available from: https://aavs.org/alternatives/research/ [Accessed: 2020-05-12]

[6] McArdle JE. The prohibition of routine ascites production of monoclonal antibodies in the United States: a successful application of the alternatives approach to biomedical research. Altern Lab Anim. 2000 Sep-Oct;28(5):725-729. doi: 10.1177/026119290002800502.

[7] Köhler G, Milstein C. Continuous cultures of fused cells secreting antibody of predefined specificity. Nature. 1975 Aug 7;256(5517):495-497. doi: 10.1038/256495a0.

[8] Parray HA, Shukla S, Samal S, et al. Hybridoma technology a versatile method for isolation of monoclonal antibodies, its applicability across species, limitations, advancement and future perspectives. Int Immunopharmacol. 2020;85:106639. doi:10.1016/j.intimp.2020.106639.

[9] Posner J, Barrington P, Brier T, Datta-Mannan A. Monoclonal Antibodies: Past, Present and Future. Handb Exp Pharmacol. 2019;260:81-141. doi: 10.1007/164_2019_323.

[10] Stanfield RL, Wilson IA. Antibody Structure. Microbiol Spectr. 2014 Apr;2(2). doi: 10.1128/microbiolspec.AID-0012-2013.

[11] Chothia C, Lesk AM. Canonical structures for the hypervariable regions of immunoglobulins. J Mol Biol. 1987 Aug 20;196(4):901-917. doi: 10.1016/0022-2836(87)90412-8.

[12] Holliger P, Hudson PJ. Engineered antibody fragments and the rise of single domains. Nat Biotechnol. 2005 Sep;23(9):1126-1136. doi: 10.1038/nbt1142.

[13] Gupta SK, Shukla P. Microbial platform technology for recombinant antibody fragment production: A review. Crit Rev Microbiol. 2017Feb;43(1):31-42. doi: 10.3109/1040841X.2016.1150959.

[14] Pothin E, Lesuisse D, Lafaye P. Brain Delivery of Single-Domain Antibodies: A Focus on VHH and VNAR. Pharmaceutics. 2020 Sep 30;12(10):937. doi: 10.3390/pharmaceutics12100937.

[15] Streltsov VA, Carmichael JA, Nuttall SD. Structure of a shark IgNAR antibody variable domain and modeling of an early-developmental isotype. Protein Sci. 2005 Nov;14(11):2901-9. doi: 10.1110/ps.051709505. Epub 2005 Sep 30. PMID: 16199666; PMCID: PMC2253229.

[16] Ronca R, Benzoni P, De Luca A, Crescini E, Dell'era P. Phage displayed

peptides/antibodies recognizing growth factors and their tyrosine kinase receptors as tools for anti-cancer therapeutics. Int J Mol Sci. 2012;13(4):5254-77. doi: 10.3390/ijms13045254. Epub 2012 Apr 24. PMID: 22606042; PMCID: PMC3344278.

[17] Rader C. Overview on concepts and applications of Fab antibody fragments. Curr Protoc Protein Sci. 2009 Feb;Chapter 6:Unit 6.9. doi: 10.1002/0471140864.ps0609s55.

[18] Kitten O, Martineau P. Les formats alternatifs aux anticorps - Fragments et nouvelles charpentes [Antibody alternative formats: antibody fragments and new frameworks]. Med Sci (Paris). 2019 Dec;35(12):1092-1097. French. doi: 10.1051/medsci/2019217.

[19] Joosten V, Lokman C, Van Den Hondel CA, Punt PJ. The production of antibody fragments and antibody fusion proteins by yeasts and filamentous fungi. Microb Cell Fact. 2003 Jan 30;2(1):1. doi: 10.1186/1475-2859-2-1.

[20] Ahmad ZA, Yeap SK, Ali AM, Ho WY, Alitheen NB, Hamid M. scFv antibody: principles and clinical application. Clin Dev Immunol. 2012;2012:980250. doi: 10.1155/2012/980250. Feb;43(1):31-42. doi: 10.3109/1040841X.2016.1150959.

[21] Kaplon H, Muralidharan M, Schneider Z, Reichert JM. Antibodies to watch in 2020. MAbs. 2020 Jan-Dec;12(1):1703531. doi: 10.1080/19420862.2019.1703531.

[22] Doerner A, Rhiel L, Zielonka S, Kolmar H. Therapeutic antibody engineering by high efficiency cell screening. FEBS Lett. 2014;588(2):278-287. Available from: http://dx.doi.org/10.1016/j.febslet.2013.11.025

[23] Plückthun A. Ribosome Display: A Perspective. In: Methods in molecular biology (Clifton, NJ). United States; 2012. p. 3-28. Available from: http://www.springerlink.com/index/10.1007/978-1-61779-379-0

[24] Chaput JC, Woodbury NW, Stearns LA, Williams BAR. Creating protein biocatalysts as tools for future industrial applications. Expert Opin Biol Ther. 2008 Aug 9;8(8):1087-1098. doi: 10.1517/14712598.8.8.1087.

[25] Barbas CF, Kang AS, Lerner RA, Benkovic SJ. Assembly of combinatorial antibody libraries on phage surfaces: the gene III site. Proc Natl Acad Sci. 1991 Sep 15;88(18):7978-7982. doi: 10.1073/pnas.88.18.7978.

[26] ørum H, Andersen PS, øster A, Johansen LK, Riise E, Bjørnvad M, et al. Efficient method for construction comprehensive murine fab antibody libraries displayed on phage. Nucleic Acids Res. 1993;21(19):4491-4498.

[27] Fuh G. Synthetic antibodies as therapeutics. Expert Opin Biol Ther. 2007 Jan 6;7(1):73-87. doi: 10.1517/14712598.7.1.73.

[28] Almagro JC, Pedraza-Escalona M, Arrieta HI, Pérez-Tapia SM. Phage Display Libraries for Antibody Therapeutic Discovery and Development. Antibodies. 2019;8(3):44.

[29] Smith GP. Filamentous fusion phage: novel expression vectors that display cloned antigens on the virion surface. Science. 1985 Jun 14;228(4705):1315-1317. Available from: https://www.sciencemag.org/lookup/doi/10.1126/science.4001944

[30] Lowman HB. Phage Display for Protein Binding. In: Encyclopedia of Biological Chemistry. 2nd ed. Elsevier; 2013. p. 431-436. Available from: http://dx.doi.org/10.1016/B978-0-12-378630-2.00061-X

[31] Wu CH, Liu IJ, Lu RM, Wu HC. Advancement and applications of

peptide phage display technology in biomedical science. J Biomed Sci. 2016;23(1):1-14. Available from: http://dx.doi.org/10.1186/s12929-016-0223-x

[32] Lu Z, Murray KS, Cleave V Van, LaVallie ER, Stahl ML, McCoy JM. Expression of Thioredoxin Random Peptide Libraries on the Escherichia coli Cell Surface as Functional Fusions to Flagellin: A System Designed for Exploring Protein-Protein Interactions. Nat Biotechnol. 1995 Apr 1;13(4):366-372. Available from: http://www.nature.com/doifinder/10.1038/nbt0495-366

[33] Derakhshandeh S, Shahrokhi N, Khalaj V, Habibi M, Moazzezy N, Asadi Karam MR, et al. Surface display of uropathogenic *Escherichia coli* FimH in Lactococcus lactis: In vitro characterization of recombinant bacteria and its protectivity in animal model. Microb Pathog. 2020 Apr;141(September 2019):103974. Available from: https://doi.org/10.1016/j.micpath.2020.103974

[34] Guo X, Cao H, Wang Y, Liu Y, Chen Y, Wang N, et al. Screening scFv antibodies against infectious bursal disease virus by co-expression of antigen and antibody in the bacteria display system. Vet Immunol Immunopathol. 2016 Nov;180:45-52. Available from: http://dx.doi.org/10.1016/j.vetimm.2016.09.004

[35] Harvey BR, Georgiou G, Hayhurst A, Jeong KJ, Iverson BL, Rogers GK. Anchored periplasmic expression, a versatile technology for the isolation of high-affinity antibodies from Escherichia coli-expressed libraries. Proc Natl Acad Sci U S A. 2004;101(25):9193-9198.

[36] Shusta E, Pepper L, Cho Y, Boder E. A Decade of Yeast Surface Display Technology: Where Are We Now? Comb Chem High Throughput Screen. 2008;11(2):127-134.

[37] Boder ET, Raeeszadeh-Sarmazdeh M, Price JV. Engineering antibodies by yeast display. Arch Biochem Biophys. 2012 Oct;526(2):99-106. Available from: http://dx.doi.org/10.1016/j.abb.2012.03.009

[38] Ho M, Pastan I. Mammalian Cell Display for Antibody Engineering. In: Dimitrov AS, editor. Methods in molecular biology. Totowa, NJ: Humana Press; 2009. p. 337-52. (Methods in Molecular BiologyTM; vol. 525). Available from: http://link.springer.com/10.1007/978-1-59745-554-1

[39] Zhou C, Jacobsen FW, Cai L, Chen Q, Shen D. Development of a novel mammalian cell surface antibody display platform. MAbs. 2010 Sep 30;2(5):508-518. Available from: http://www.tandfonline.com/doi/abs/10.4161/mabs.2.5.12970

[40] Tomimatsu K, Matsumoto SE, Tanaka H, Yamashita M, Nakanishi H, Teruya K, et al. A rapid screening and production method using a novel mammalian cell display to isolate human monoclonal antibodies. Biochem Biophys Res Commun. 2013;441(1):59-64. doi: 10.1016/j.bbrc.2013.10.007.

[41] Newton MS, Cabezas-Perusse Y, Tong CL, Seelig B. In Vitro Selection of Peptides and Proteins-Advantages of mRNA Display. ACS Synth Biol. 2020;9(2):181-190.

[42] Lipovsek D, Plückthun A. In-vitro protein evolution by ribosome display and mRNA display. J Immunol Methods. 2004 Jul;290(1-2):51-67. doi: 10.1016/j.jim.2004.04.008. Erratum in: J Immunol Methods. 2004 Nov;294(1-2):213.

[43] Seelig B. mRNA display for the selection and evolution of enzymes from in vitro-translated protein libraries. Nat Protoc. 2011;6(4):540-552. Available from: http://dx.doi.org/10.1038/nprot.2011.312

[44] Krawczyk K, Dunbar J, Deane CM. Computational Tools for Aiding Rational Antibody Design. Methods Mol Biol. 2017;1529:399-416. doi: 10.1007/978-1-4939-6637-0_21.

[45] Brenke R, Hall DR, Chuang GY, Comeau SR, Bohnuud T, Beglov D, Schueler-Furman O, Vajda S, Kozakov D. Application of asymmetric statistical potentials to antibody-protein docking. Bioinformatics. 2012 Oct 15;28(20):2608-2614. doi: 10.1093/bioinformatics/bts493.

[46] Esmaielbeiki R, Krawczyk K, Knapp B, Nebel JC, Deane CM. Progress and challenges in predicting protein interfaces. Brief Bioinform. 2016 Jan;17(1):117-131. doi: 10.1093/bib/bbv027.

[47] Marcatili P, Rosi A, Tramontano A. PIGS: automatic prediction of antibody structures. Bioinformatics. 2008 Sep 1;24(17):1953-1954. doi: 10.1093/bioinformatics/btn341.

[48] Sircar A, Kim ET, Gray JJ. RosettaAntibody: antibody variable region homology modeling server. Nucleic Acids Res. 2009 Jul;37(Web Server issue):W474–W479. doi: 10.1093/nar/gkp387.

[49] Whitelegg NR, Rees AR. WAM: an improved algorithm for modelling antibodies on the WEB. Protein Eng. 2000 Dec;13(12):819-824. doi: 10.1093/protein/13.12.819.

[50] Almagro JC, Teplyakov A, Luo J, Sweet RW, Kodangattil S, Hernandez-Guzman F, Gilliland GL. Second antibody modeling assessment (AMA-II). Proteins. 2014 Aug;82(8):1553-1562. doi: 10.1002/prot.24567.

[51] Regep C, Georges G, Shi J, Popovic B, Deane CM. The H3 loop of antibodies shows unique structural characteristics. Proteins. 2017 Jul;85(7):1311-1318. doi: 10.1002/prot.25291.

[52] Marks C, Deane CM. Antibody H3 Structure Prediction. Comput Struct Biotechnol J. 2017 Feb 1;15:222-231. doi: 10.1016/j.csbj.2017.01.010.

[53] Cho KI, Lee K, Lee KH, Kim D, Lee D. Specificity of molecular interactions in transient protein-protein interaction interfaces. Proteins. 2006 Nov 15;65(3):593-606. doi: 10.1002/prot.21056.

[54] Zacharias M. Accounting for conformational changes during protein-protein docking. Curr Opin Struct Biol. 2010 Apr;20(2):180-186. doi: 10.1016/j.sbi.2010.02.001.

[55] DeKosky BJ, Lungu OI, Park D, Johnson EL, Charab W, Chrysostomou C, Kuroda D, Ellington AD, Ippolito GC, Gray JJ, Georgiou G. Large-scale sequence and structural comparisons of human naive and antigen-experienced antibody repertoires. Proc Natl Acad Sci U S A. 2016 May 10;113(19):E2636–E2645. doi: 10.1073/pnas.1525510113.

[56] Nimrod G, Fischman S, Austin M, Herman A, Keyes F, Leiderman O, Hargreaves D, Strajbl M, Breed J, Klompus S, Minton K, Spooner J, Buchanan A, Vaughan TJ, Ofran Y. Computational Design of Epitope-Specific Functional Antibodies. Cell Rep. 2018 Nov 20;25(8):2121-2131.e5. doi: 10.1016/j.celrep.2018.10.081.

[57] Clavero-Álvarez A, Di Mambro T, Perez-Gaviro S, Magnani M, Bruscolini P. Humanization of Antibodies using a Statistical Inference Approach. Sci Rep. 2018 Oct 4;8(1):14820. doi: 10.1038/s41598-018-32986-y.

[58] Kurella VB, Gali R. Antibody Design and Humanization via In Silico Modeling. Methods Mol Biol. 2018;1827:3-14. doi: 10.1007/978-1-4939-8648-4_1.

[59] Vazquez-Lombardi R, Nevoltris D, Luthra A, Schofield P, Zimmermann C, Christ D. Transient expression of human antibodies in mammalian cells. Nat Protoc. 2018 Jan;13(1):99-117. doi: 10.1038/nprot.2017.126.

[60] Beck A, Wagner-Rousset E, Bussat MC, Lokteff M, Klinguer-Hamour C, Haeuw JF, Goetsch L, Wurch T, Van Dorsselaer A, Corvaïa N. Trends in glycosylation, glycoanalysis and glycoengineering of therapeutic antibodies and Fc-fusion proteins. Curr Pharm Biotechnol. 2008 Dec;9(6):482-501. doi: 10.2174/138920108786786411.

[61] Kunert R, Reinhart D. Advances in recombinant antibody manufacturing. Appl Microbiol Biotechnol. 2016 Apr;100(8):3451-3461. doi: 10.1007/s00253-016-7388-9.

[62] Pham, P (2018). Medical Biotechnology: Techniques and applications. In: Barh, D; Azevedo V, editors. Omics Technologies and Bio-Engineering. Vol. 1. 1st ed. PLACE: Academic Press; 2018. p. 449-469. doi:10.1016/b978-0-12-804659-3.00019-1

[63] Lu RM, Hwang YC, Liu IJ, Lee CC, Tsai HZ, Li HJ, Wu HC. Development of therapeutic antibodies for the treatment of diseases. J Biomed Sci. 2020 Jan 2;27(1):1. doi: 10.1186/s12929-019-0592-z.

[64] Animal Cell Technology Industrial Platform. ACTIP: Monoclonal Antibodies Approved by the EMA and FDA for Therapeutic Use (status 2017). [Internet].2020. Available from: https://www.actip.org/products/monoclonal-antibodies-approved-by-the-ema-and-fda-for-therapeutic-use/ [Accessed: 2020-12-09]

[65] Capela, E. V., Aires-Barros, M. R., Freire, M. G., & Azevedo, A. M. (2017). Monoclonal antibodies—addressing the challenges on the manufacturing processing of an advanced class of therapeutic agents. Front Clini Drug Res Anti Infect, 4(4), 142.

[66] Taki S, Kamada H, Inoue M, Nagano K, Mukai Y, Higashisaka K, Yoshioka Y, Tsutsumi Y, Tsunoda S. A Novel Bispecific Antibody against Human CD3 and Ephrin Receptor A10 for Breast Cancer Therapy. PLoS One. 2015 Dec 17;10(12):e0144712. doi: 10.1371/journal.pone.0144712.

[67] Zhong X, Ma W, Meade CL, Tam AS, Llewellyn E, Cornell R, Cote K, Scarcelli JJ, Marshall JK, Tzvetkova B, Figueroa B, DiNino D, Sievers A, Lee C, Guo J, Mahan E, Francis C, Lam K, D'Antona AM, Zollner R, Zhu HL, Kriz R, Somers W, Lin L. Transient CHO expression platform for robust antibody production and its enhanced N-glycan sialylation on therapeutic glycoproteins. Biotechnol Prog. 2019 Jan;35(1):e2724. doi: 10.1002/btpr.2724.

[68] Hunter M, Yuan P, Vavilala D, Fox M. Optimization of Protein Expression in Mammalian Cells. Curr Protoc Protein Sci. 2019 Feb;95(1):e77. doi: 10.1002/cpps.77.

[69] Liu H, Liu XM, Li SC, Wu BC, Ye LL, Wang QW, Chen ZL. A high-yield and scaleable adenovirus vector production process based on high density perfusion culture of HEK 293 cells as suspended aggregates. J Biosci Bioeng. 2009 May;107(5):524-529. doi: 10.1016/j.jbiosc.2009.01.004.

[70] Liu XM, Liu H, Wu BC, Li SC, Ye LL, Wang QW, Huang PT, Chen ZL. Suspended aggregates as an immobilization mode for high-density perfusion culture of HEK 293 cells in a stirred tank bioreactor. Appl Microbiol Biotechnol. 2006 Oct;72(6):1144-1151. doi: 10.1007/s00253-006-0409-3.

[71] Jones D, Kroos N, Anema R, van Montfort B, Vooys A, van der Kraats S, van der Helm E, Smits S, Schouten J, Brouwer K, Lagerwerf F, van Berkel P,

Opstelten DJ, Logtenberg T, Bout A. High-level expression of recombinant IgG in the human cell line per.c6. Biotechnol Prog. 2003 Jan-Feb;19(1):163-168. doi: 10.1021/bp025574h.

[72] Tsuruta LR, Lopes Dos Santos M, Yeda FP, Okamoto OK, Moro AM. Genetic analyses of Per.C6 cell clones producing a therapeutic monoclonal antibody regarding productivity and long-term stability. Appl Microbiol Biotechnol. 2016 Dec;100(23):10031-10041. doi: 10.1007/s00253-016-7841-9. Epub 2016 Sep 27.

[73] Frenzel A, Hust M, Schirrmann T. Expression of recombinant antibodies. Front Immunol. 2013 Jul 29;4:217. doi: 10.3389/fimmu.2013.00217.

[74] Stephenson KE, Keefer MC, Bunce CA, Frances D, Abbink P, Maxfield LF, Neubauer GH, Nkolola J, Peter L, Lane C, Park H, Verlinde C, Lombardo A, Yallop C, Havenga M, Fast P, Treanor J, Barouch DH. First-in-human randomized controlled trial of an oral, replicating adenovirus 26 vector vaccine for HIV-1. PLoS One. 2018 Nov 14;13(11):e0205139. doi: 10.1371/journal.pone.0205139.

[75] Bockstal V, Tiemessen MM, Achterberg R, Van Wordragen C, Knaapen AM, Serroyen J, Marissen WE, Schuitemaker H, Zahn R. An inactivated poliovirus vaccine using Sabin strains produced on the serum-free PER.C6® cell culture platform is immunogenic and safe in a non-human primate model. Vaccine. 2018 Nov 12;36(46):6979-6987. doi: 10.1016/j.vaccine.2018.09.068.

[76] Lindegren S, Andrade LN, Bäck T, Machado CM, Horta BB, Buchpiguel C, Moro AM, Okamoto OK, Jacobsson L, Cederkrantz E, Washiyama K, Aneheim E, Palm S, Jensen H, Tuma MC, Chammas R, Hultborn R, Albertsson P. Binding Affinity, Specificity and Comparative Biodistribution of the Parental Murine Monoclonal Antibody MX35 (Anti-NaPi2b) and Its Humanized Version Rebmab200. PLoS One. 2015 May 13;10(5):e0126298. doi: 10.1371/journal.pone.0126298.

[77] Anywaine Z, Whitworth H, Kaleebu P, Praygod G, Shukarev G, Manno D, Kapiga S, Grosskurth H, Kalluvya S, Bockstal V, Anumendem D, Luhn K, Robinson C, Douoguih M, Watson-Jones D. Safety and Immunogenicity of a 2-Dose Heterologous Vaccination Regimen With Ad26.ZEBOV and MVA-BN-Filo Ebola Vaccines: 12-Month Data From a Phase 1 Randomized Clinical Trial in Uganda and Tanzania. J Infect Dis. 2019 Jun 5;220(1):46-56. doi: 10.1093/infdis/jiz070.

[78] Baltz, R, Demain, A, & Davies, J, editors. Manual of industrial microbiology and biotechnology. 3rd ed. Washington: American Society for Microbiology Press; 2010. 784 p.

[79] Dhara VG, Naik HM, Majewska NI, Betenbaugh MJ. Recombinant Antibody Production in CHO and NS0 Cells: Differences and Similarities. BioDrugs. 2018 Dec;32(6):571-584. doi: 10.1007/s40259-018-0319-9.

[80] Gutiérrez-Granados S, Cervera L, Kamen AA, Gòdia F. Advancements in mammalian cell transient gene expression (TGE) technology for accelerated production of biologics. Crit Rev Biotechnol. 2018 Sep;38(6):918-940. doi: 10.1080/07388551.2017.1419459.

[81] Liu CH, Liu YX, Wu WC. Facile development of medium optimization for antibody production: implementation in spinner flask and hollow fiber reactor. Cytotechnology. 2018 Dec;70(6):1631-1642. doi: 10.1007/s10616-018-0255-z.

[82] Phelan K, May KM. Basic Techniques in Mammalian Cell Tissue Culture. Curr

Protoc Toxicol. 2016 Nov 1;70:A.3B.1-A.3B.22. doi: 10.1002/cptx.13.

[83] Berson RE, Friederichs G. A self-feeding roller bottle for continuous cell culture. Biotechnol Prog. 2008 Jan-Feb;24(1):154-157. doi: 10.1021/bp070122w.

[84] Buchsteiner M, Quek LE, Gray P, Nielsen LK. Improving culture performance and antibody production in CHO cell culture processes by reducing the Warburg effect. Biotechnol Bioeng. 2018 Sep;115(9):2315-2327. doi: 10.1002/bit.26724.

[85] Li F, Vijayasankaran N, Shen AY, Kiss R, Amanullah A. Cell culture processes for monoclonal antibody production. MAbs. 2010;2(5):466-479. doi:10.4161/mabs.2.5.12720

[86] Costa, A.R.; Nóbrega, F.L.; Santos, S.B.; Kluskens, L.D.; Azeredo, J. Molecular recognition in host-pathogen interactions. The 5th EMBO Meeting: Advancing Life Sciences. Amsterdam, Netherlands, 21-24 September, 30, pp. 563-572. 2013;

[87] Lucas Silva Carvalho, Otávio Bravim da Silva, Gabriela Carneiro de Almeida, Juliana Davies de Oliveira, Nadia Skorupa Parachin and Talita Souza Carmo (February 8th 2017). Production Processes for Monoclonal Antibodies, Fermentation Processes, Angela Faustino Jozala, IntechOpen, DOI: 10.5772/64263.

[88] Walsh G. Biopharmaceutical benchmarks 2010. Nat Biotechnol. 2010 Sep;28(9):917-924. doi: 10.1038/nbt0910-917.

[89] Chelliapan, S. and Sallis, P.J. (2013) Removal of organic compound from pharmaceutical wastewater using advanced oxidation processes. J. Sci. Ind. Res. 72, 248-254.

[90] Vincent Sewalt, Diane Shanahan, Lori Gregg, James La Marta, and Roberto Carrillo.Industrial Biotechnology.Oct 2016.295-302.http://doi.org/10.1089/ind.2016.0011

[91] Lee, Y.J., Jeong, K.J., 2015. Challenges to production of antibodies in bacteria and yeast. J. Biosci. Bioeng. 120, 483-490.

[92] Darby, R.A.J., S.P. Cartwright, M.V. Dilworth, and R.M. Bill, Which Yeast Species Shall I Choose? Saccharomyces cerevisiae Versus Pichia pastoris (Review), in Recombinant Protein Production in Yeast: Methods and Protocols, M.R. Bill, Editor. 2012, Humana Press: Totowa, NJ. p. 11-23.

[93] Zalai, D. et al. (2012) A dynamic fed batch strategy for a Pichia pastoris mixed feed system to increase process understanding. Biotechnol. Prog. 28, 878-886.

[94] Jung, S. T., & Kim, D.-I. (2018). Mass Production of Full-Length IgG Monoclonal Antibodies from Mammalian, Yeast, and Bacterial Hosts. Emerging Areas in Bioengineering, 679-695. doi:10.1002/9783527803293.ch39

[95] Robinson MP, Ke N, Lobstein J, Peterson C, Szkodny A, Mansell TJ, Tuckey C, Riggs PD, Colussi PA, Noren CJ, Taron CH, DeLisa MP, Berkmen M. Efficient expression of full-length antibodies in the cytoplasm of engineered bacteria. Nat Commun. 2015 Aug 27;6:8072. doi: 10.1038/ncomms9072.

[96] Shoeb E, Hefferon K. Future of cancer immunotherapy using plant virus-based nanoparticles. Future Sci OA. 2019 Jul 25;5(7):FSO401. doi: 10.2144/fsoa-2019-0001.

[97] Moussavou G, Ko K, Lee JH, Choo YK. Production of monoclonal antibodies in plants for cancer immunotherapy. Biomed Res Int. 2015;2015:306164. doi: 10.1155/2015/306164

[98] Nessa MU, Rahman MA, Kabir Y. Plant-Produced Monoclonal Antibody as Immunotherapy for Cancer. Biomed Res Int. 2020 Aug 24;2020:3038564. doi: 10.1155/2020/3038564.

[99] Bosch D, Castilho A, Loos A, Schots A, Steinkellner H. N-glycosylation of plant-produced recombinant proteins. Curr Pharm Des. 2013;19(31):5503-5512. doi: 10.2174/1381612811319310006.

[100] Jin C, Altmann F, Strasser R, Mach L, Schähs M, Kunert R, Rademacher T, Glössl J, Steinkellner H. A plant-derived human monoclonal antibody induces an anti-carbohydrate immune response in rabbits. Glycobiology. 2008 Mar;18(3):235-241. doi: 10.1093/glycob/cwm137.

[101] Strasser R, Altmann F, Steinkellner H. Controlled glycosylation of plant-produced recombinant proteins. Curr Opin Biotechnol. 2014 Dec;30:95-100. doi: 10.1016/j.copbio.2014.06.008.

[102] Castilho A, Steinkellner H. Glyco-engineering in plants to produce human-like N-glycan structures. Biotechnol J. 2012 Sep;7(9):1088-1098. doi: 10.1002/biot.201200032.

[103] Donini M, Marusic C. Current state-of-the-art in plant-based antibody production systems. Biotechnol Lett. 2019 Mar;41(3):335-346. doi: 10.1007/s10529-019-02651-z.

[104] Hiatt A, Cafferkey R, Bowdish K. Production of antibodies in transgenic plants. Nature. 1989 Nov 2;342(6245):76-78. doi: 10.1038/342076a0.

[105] Mor TS. Molecular pharming's foot in the FDA's door: Protalix's trailblazing story. Biotechnol Lett. 2015 Nov;37(11):2147-2150. doi: 10.1007/s10529-015-1908-z.

[106] Clinical Trials. Available from: https://clinicaltrials.gov/ct2/show/NCT03301051?term=medicago [Accessed: 2020-11-09]

[107] Lim S, Kim DS, Ko K. Expression of a Large Single-Chain 13F6 Antibody with Binding Activity against Ebola Virus-Like Particles in a Plant System. Int J Mol Sci. 2020 Sep. 23;21(19):7007. doi: 10.3390/ijms21197007.

[108] Zahara K, Bibi Y, Ajmal M, Sadaf HM, Bibi F, Sardar N, Laraib S, Riaz I. Tobacco plant: A possible key to Ebola vaccine. Jclm. 2017 May 5(5) 206:211 doi: /10.12980/jclm.5.2017J7-9.

Chapter 3

Monospecific and Polyreactive Monoclonal Antibodies against Human Leukocyte Antigen-E: Diagnostic and Therapeutic Relevance

Mepur H. Ravindranath and Fatiha E.L. Hilali

Abstract

A monoclonal antibody (mAb) binds to an antigen recognizing an epitope (a sequence of amino acids). A protein antigen may carry amino acid sequence unique to that antigen as well as sequences found in other proteins. Human leukocyte antigens (HLA), a family of proteins expressed by the Major Histocompatibility Complex gene family represent a special case, in that it displays a high degree of polymorphism. Every HLA molecule possesses both specific (private) epitopes and epitopes shared (public) with other HLA class Ia and class Ib molecules. HLA-E is overexpressed in cancer cells more than any other HLA Class I molecules. Therefore specific localization of HLA-E with mAbs is pivotal for developing targeted therapy against cancer. However, the commercially available mAbs for immunodiagnosis are polyreactive. We have developed anti-HLA-E mAbs and distinguished monospecific from polyreactive mAbs using Luminex multiplex single antigen bead (SAB) assay. HLA-E-binding of monospecific-mAbs was also inhibited by E-restricted epitopes. The amino acid sequences in the region of the epitopes bind to CD94/NKG2A receptors on CD8+ T cells and NK cells and block their antitumor functions. Monospecific-HLA-E mAbs recognizing the epitopes sequences can interfere with the binding to restore the anti-tumor efficacy of NK cells. Also, monospecific-mAbs augment the proliferation of CD4-/CD+ cytotoxic T-lymphocytes. Therefore, anti-HLA-E monospecific-mAb can serve as a double-edged sword for eliminating tumor cells.

Keywords: human leukocyte antigen (HLA), epitope, monospecific, polyreactive, cytotoxic T-lymphocytes, inhibitory receptors, NK cells

1. Introduction

An in-depth understanding of amino acid sequences and conformations of primary antigens recognized by any monoclonal antibody (mAb) is a necessary prerequisite for clarifying the specificity and functional limitations of a mAb. A protein antigen may be glycosylated or can occur as a monomer or a dimer or a trimer.

In this regard, human leukocyte antigen (HLA) classes are a structurally identical complex family of glycosylated homo- or hetero-dimeric proteins. They are expressed on cell surface complexed with an exogenous or endogenous peptide, as trimers. Defining the monospecificity of mAb raised against one family member of HLA is challenging. Often anti-HLA mAbs are polyreactive in that they bind to sequences common to all family member antigens, which are also known as "public epitopes". It is difficult to identify mAbs binding to unique sequences or private epitopes. Identifying such monospecific mAbs are critical for defining specific functions of antigens. Although sensitive and specific assay protocols are available to define the monospecificity of mAbs, many commercial mAbs, apparently specific for a unique HLA antigen, remain without defining their monospecificity. This review aims to distinguish monospecific mAbs that recognize private epitopes from polyreactive mAbs that bind to public epitopes of one of the HLA class Ib molecules, namely HLA-E, commonly overexpressed on human cancers. A pool of mouse mAbs was developed at Terasaki Foundation Laboratory (TFL) after immunizing with HLA-E. After validating the monospecificity of anti-HLA-E mAbs, their diagnostic and therapeutic potentials have been evaluated. These include (i) immunolocalization of cell surface expression HLA-E on human cancers, (ii) upregulation of CD8+ cytotoxic T lymphocytes, and (iii) restoration of antitumor activity of CD8+ T cells, NKT cells, and NK cells by preventing binding of HLA-E expressed on cancer cells to the inhibitory receptors (CD94/NKG2A) on the immune cells.

2. Nature and characteristics of human leukocyte antigens

Human Leukocyte antigens (HLA) are a subgroup of the Major Histocompatibility Complex (MHC) gene family. The genes that encode the HLA class-I and class-II antigens are located on the short arm of human chromosome 6 [1]. Three constituent regions of the HLA gene complex are illustrated in **Figure 1**. Class, I genes are those encoding the heavy chains (HC) or α chains, of the six class I isoforms HLA-A, -B, -C, -E, -F, and -G. Extensive polymorphism of the glycosylated heavy chains of these HLA molecules are presented in **Table 1**. We carry a pair of alleles that represent each isoform derived from their mother and father (**Table 2**). Understanding HLA profiles of a patient is necessary when administering mAbs targeting a particular HLA molecule, for amino acid sequences of target HLA may cross-react with other HLA alleles of the patient. Native HLA-I

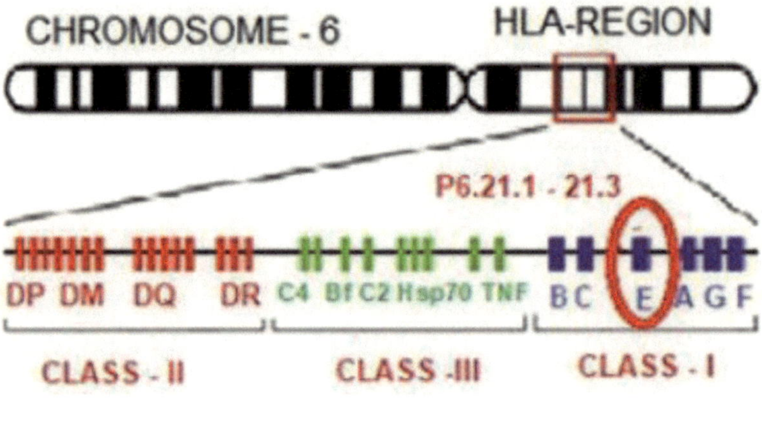

Figure 1.
Profile of the HLA gene complex on chromosome 6. All regions contain additional genes.

HLA Class I						
Gene	*A*	*B*	*C*	*E*	*F*	*G*
Alleles	6,291	7,582	6,223	256	45	82
Proteins	3,896	4,803	3,681	110	6	22

Table 1.
Numbers of HLA alleles (as of September 2020) and their proteins. See updated information at https://www.ebi.ac.uk/ipd/imgt/hla/stats.html.

PROFILES OF HLA TYPING: HLA ISOFORMS AND THEIR ALLELES					
HLA CLASS	ISOFORMS	BROTHER*		SISTER	
I	A*	**[11:02]**	[33:01]	[01:01]	**[11:02]**
I	B*	[15:01]	[58:01]	[40:01]	[57:01]
I	C*	[15:02]	[15:02]	[03:04]	[06:02]
II	DRB1	[04:03]	[13:02]	[07:01]	[11:01]
II	DRB3,4,5	[3*03:01]	**[4*01:01]**	[3*02:02]	**[4*01:01]**
II	DQA	**[01:02]**	**[03:01]**	**[01:02]**	**[03:01]**
II	DQB	[03:01]	[06:09]	[02:02]	**[03:01]**
II	DPA	**[01:03]**	**[01:03]**	**[01:03]**	**[02:01]**
II	DPB	[02:01]	[03:01]	[01:07]	[01:11]

**Mepur H. Ravindranath (brother) and his first sister.*
The alleles in bold letters refer to alleles shared by the brother and the sister.

Table 2.
Pair of HLA alleles representing each of the commonly typed HLA isoforms.

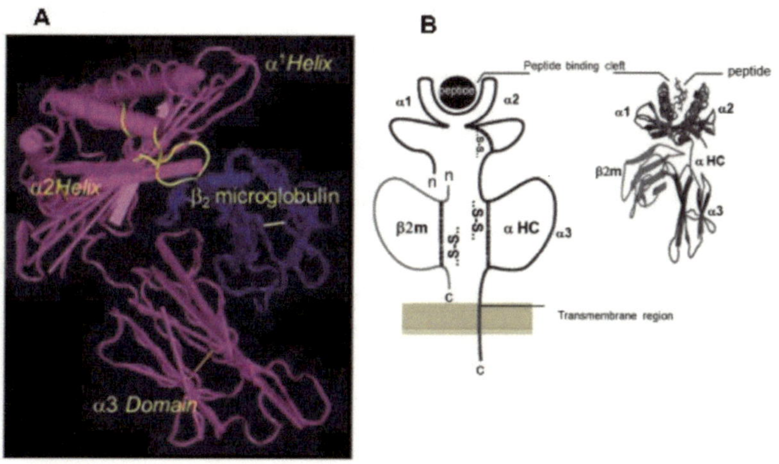

Figure 2.
(A) Conformational structure of HLA class I. the native HLA-I proteins are expressed on the cell surface as hetero-dimers, the heavy chain in combination with β2-microglobulin (β2-m). (B) the hetero-dimer on the cell surface may carry a short peptide to generate trimeric structure, designated as "closed conformer"(CC).

proteins are expressed on the cell surface as hetero-dimers, in combination with β2-microglobulin (β2-m) (**Figure 2A**). The gene encoding β2-m is situated on human chromosome 15. The hetero-dimers may also carry a peptide to form a trimer (**Figure 2B**), which is designated as "Closed Conformers (CCs)" [2]. Under the influence of cytokines (e.g. IFN- ɣ) and other activating factors (e.g. T-cell

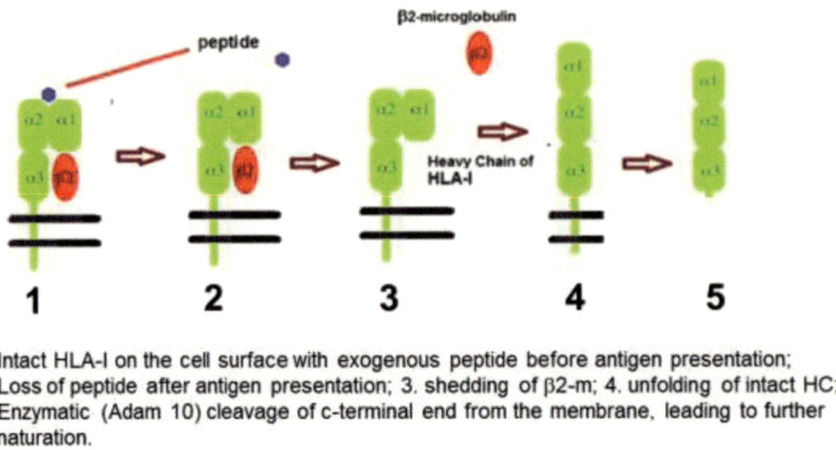

Figure 3.
The fate of HLA-I molecule after antigen presentation.

antibodies) or during inflammation, infection and tumorigenesis, the surface of metabolically active cells express only monomeric HLA heavy chains, called "Open Conformers (OCs) [3]. The examples include human T-lymphocytes activated *in vitro* and *in vivo,* as well as by EBV-transformed B-cells, CD19+ B-cells, CD8+ T cells, CD56+ NK-cells, CD14+ monocytes, extravillous trophoblasts and monocytes, dendritic cells (DCs), B-cell lines (RAJI, NALM6), and the myeloid cell line (KG-1A) [4–12]. The kinetics of conformational alterations in the naturally-occurring HLA-I OCs after activation has been investigated in healthy human T-cells [11]. The cytoplasmic c-terminal tail of naturally-occurring HLA-I OCs is tyrosine phosphorylated and plays a role in signal transduction [11].

HLA-I on antigen-presenting cells presents endogenous (intracellular) peptides. Importantly, viral peptides that have been broken by the proteasome are transferred to the endoplasmic reticulum (ER) via transporters (TAP). In ER, peptides are processed with OCs of HLA-I and exported to the cell surface as a trimer for presentation to T-cell receptors of CD8+ T-cells. This strategy kills the cell, thus preventing viral replication. After antigen presentation, the HLA-I is degraded (**Figure 3**). Ultimately, such degradation results in exposing the cryptic epitopes on the OCs to an individual's own immune system. Antibodies formed against the cryptic epitopes eliminate the degraded HLA from the circulation. The antibody-producing cells may remain hidden and silent for long periods. They are referred to as "long-lived B cells" [13]. Evidently, anti-HLA antibodies occur in normal and healthy individuals [14–16], as well as in the pooled and purified plasma also known as intravenous immunoglobulin (IVIg) [16, 17].

3. Diagnostic and clinical relevance of non-classical HLA class Ib antigens

Unlike classical HLA-Ia (HLA-A, HLA-B & HLA-C), non-classical HLA-Ib (HLA-E, HLA-F & HLA-G) genes and molecules are oligotrophic, with restricted and selective tissue distribution [18–20]. HLA-Ib molecules are expressed in a diverse array of cells including T and B lymphocytes, Natural Killer Cells, monocytes, macrophages, megakaryocytes, and organs i.e., lymph nodes, spleen, skin, salivary glands, thyroid, stomach, liver, kidney, urinary bladder, endometrial, and trophoblasts. Their overexpression is reported on activated T cells bone marrow cells inflamed cells and tissues (e.g. synovial fibroblasts), tumor cells [21–24].

The HLA-Ib molecules are capable of interacting with cell-surface receptors present on specific immune-cell subsets, inducing activation or inhibition of signaling cascades within such specific immune cells as NK cells, macrophages, and dendritic cells [25–27]. Their interaction with different immunomodulatory (activating and/or inhibiting) cell-surface receptors on NK cells and macrophages signify their role in innate immunity; these receptors include CD94/NKG2, Ig-like transcript 2 (ILT2), Ig-like transcript 4 (ILT4), KIR2DL4, and CD160. These interactions are a component of innate immunity [27]; e.g., HLA-Ib is expressed during pregnancy, playing a major role in tolerance shown towards the fetus and placenta [28–34]. HLA-Ib molecules also generate a pool of antibodies *in vivo*, which may include monospecific or polyreactive (cross-reactive with other HLA-I molecule [16, 35–39]. Soluble HLA-Ib is also found in the synovial fluid and the circulation of healthy and in cancer patients [40–42].

4. Human leukocyte antigen-E (HLA-E)

4.1 Unique characteristics of HLA-E

Although several alleles of HLA-E (**Table 1**) exist, only two are extensively distributed among different ethnic groups [43]. The alleles differ by a single amino acid at position 107 [44–46]; Arginine in HLA-E^{R107} (HLAE*01:01) is replaced by glycine in HLA-E^{G107} (HLA-E*03:01) [45]. Such amino acid substitution influence thermal stability, which results in a more stable expression of cell surface HLA-E*01:03 compared to HLA-E*01:01 [44], including half-life of the molecule. HLA-E*01:01 and HLA-E*03:01 bind to different restricted sets of peptides.

HLA-E present peptides derived from HLA-Ia signal sequences (leader peptides), heat-shock protein (Hsp-60), human cytomegalovirus, Hepatitis C virus, Human Immunodeficiency Virus, Epstein Barr virus, Influenza virus, *Salmonella enteric* and *Mycobacterium* glycoproteins to T-lymphocytes [46–49]. The binding of HLA-E to the leader peptides of HLA-Ia stabilizes the HLA-E and enables migration to the cell surface [49]. When HLA-E does not reach the cell surface of a tumor cell, the cell is susceptible to lysis by NK cells. The crystallographic analyses of HLA-E structure reveals the molecular mechanisms underlying this function of HLA-E [24]. Importantly, tumor-associated HLA-E can be shed into the tumor microenvironment and circulation as soluble HLA-E (sHLA-E) [23, 50–56].

4.2 HLA-E expression on cancer cells using mAb-based diagnostic assays: Limitations and reliability

The literature (**Table 3**) on HLA-E expression on human cancers based on the commercially available diagnostic anti-HLA-E mAbs tests, reveals that none of the diagnostic mAbs were tested for their unique or monospecificity for HLA-E. If the mAb is not specific for the unique epitopes of antigen and if it binds to public epitopes or epitopes shared by a family of antigens, then data is unjustified to conclude the expression HLA-E. Principally this criterion is valid for any diagnostic or therapeutic antibody. We have undertaken efforts to examine, using Luminex multiplex SAB assay, the specificity of commercial anti-HLA-E mAbs employed in the 47 clinical studies (**Table 3**). Summary of the results [16, 21, 35–39, 96–98] is

NATURE OF HUMAN CANCER	COMMERCIAL mAbs	REFERENCES
Melanoma Cervical Cancer	3D12	Marín R et al. Immunogenetics. 54(11):767–75.2003 [57]
Melanoma	MEM-E/02	Derré L et al. J Immunol. 177:3100–7. 2006. [22]
Melanoma and other cancers	MEM-E/07	Allard M et al. PLoS One 6(6):e21118, 2011 [55]
	MEM-E/08	
Lip squamousal cell carcinoma	MEM-E/02	Goncalves et al. Human Immunol. 77(9): 785–790, 2016 [58]
Laryngeal carcinoma	MEM-E/02	Silva TG et al. Histol Histopathol. 26:1487–97. 2011 [59]
Vulvar intraepithelial carcinoma	MEM-E/02	van Esch EM et al. Int J Cancer. 135(4): 830–42, 2014 [60]
Penile Cancer	MEM-E/02	Djajadiningrat et al. J Urol. 193(4):1245–51. 2015. [61]
Glioblastomas	MEM-E/02	Mittelbronn, M. et al., J. Neuroimmunol. 189: 50–58. 2007 [62]
Glioblastomas	MEM-E/02	Kren L et al. J Neuroimmunol. 220:131–5. 2010 [63]
Glioblastomas	MEM-E/02	Kren L et al. Neuropathology. 31: 129–34. 2011 [64]
Glioblastomas stem cells	3D12	Wolpert et al. J Neuroimmunol. 250(1–2):27–34 2012 [65]
Glioblastomas	3D12	Wischhusen J et al. J Neuropathol Exp Neurol. 64:523–8. 2005 [66]
Neuroblastoma	3H2679	Zhen et al. Oncotarget. 7(28): 44340–44349, 2016. [67]
Neuroblastoma	3D12	Morandi et al. J Immunol Res. 2016:7465741, 2016. [53]
Oral Osteosarcoma	MEM-E/02	Costa Arantes et al. Oral Surg Oral Med Oral Pathol Oral Radiol. 123(6):e188-e196. 2017. [68]
Intraoral mucoepidermoid carcinoma	MEM-E/02	Mosconi C Arch Oral Biol. 83:55–62, 2017. [69]
Rectal Cancer	MEM-E/02	Reimers et al. BMC Cancer BMC Cancer. 14:486.1–12, 2014. [70]
Colorectal carcinoma	MEM-E/08	Levy et al. Int J Oncol. 32(3): 633–41. 2008 [71]
Colorectal carcinoma	MEM-E/08	Levy et al. Innate Immun. 15(2):91–100. 2009. [72]
Colorectal carcinoma	MEM-E/02	Benevolo M, et al. J Transl Med. 9:184. 2011. [73]
Colorectal carcinoma	MEM-E/02?	Bossard C et al. Int J Cancer. 131 (4): 855–863. 2012. [67]
Colorectal carcinoma	MEM-E/02?	Zhen et al., Med Oncol. 30(1):482. 2013. [74]
Colorectal carcinoma	MEM-E/02	Zeestraten et al. Br J Cancer. 110(2):459–68. 2014. [75]
Colorectal carcinoma	MEM-E/02	Guo et al. Cell Immunol. 293(1):10–6, 2015. [76]
Colorectal carcinoma	3H2679	Ozgul Ozdemir et al. Ann Diagn Pathol. 25:60–63, 2016 [77]
Colorectal carcinoma	MEM-E/02	Huang et al. Oncol Lett. 13(5):3379–3386, 2017. [78]
Colon carcinoma and leukemia (K562)	MEM-E/06	Stangl S et al. Cell Stress Chaperones. 13(2):221–30. 2008. [79]
Colon carcinoma	MEM-E/02	Zeestraten EC et al. Br J Cancer. 110(2): 459–68.2014. [75]
Hepatocellular carcinoma	MEM-E/02	Chen et al. Neoplasma. 58(5):371–376, 2011. [80]
Non-small cell Lung Carcinoma	MEM-E/02	Talebian-Yazdi et al. Oncotarget. 7(3):3477–3488, 2016. [81]

NATURE OF HUMAN CANCER	COMMERCIAL mAbs	REFERENCES
Breast cancer	MEM-E/02	de Kruijf EM et al. J Immunol. 185:7452, 2010 [82]
Breast cancer	MEM-E/02	da Silva et al. Int J Breast Cancer. 2013:250435. 2013. [83]
Ovarian cancer/ Cervical cancer	MEM-E/02	Gooden M et al.PNAS USA 108:10656, 2011. [84]
Cervical cancer	MEM-E/02	Gonçalves MA et al. Eur J Obstet Gynecol Reprod Biol. 141:70–4. 2008. [85]
Cervical cancer	MEM-E/02	Spaans VM et al., J Transl Med. 10:184. 2012. [86]
Cervical squamous and adenocarcinoma	MEM-E/02	Ferns et al. J Immunother Cancer. 4:78, 2016. [87]
Serous Ovarian Adenocarcinoma	MEM-E/02	Andersson et al. Oncoimmunology, 25;5(1):e1052213, 2015. [88]
Serous Ovarian Adenocarcinoma	MEM-E/02	Zheng et al. Cancer Sci. 106(5): 522–528, 2015. [89]
Renal Cell Carcinoma	MEM-E/02	Hanak L et al. Med Sci Monit. 15(12):CR638–43.2009. [90]
Renal Cell Carcinoma	MEM-E/02	Kren L et al., Diagnostic Pathology, 7:58, 2012 [91]
Thyroid cancer	MEM-E/02	Zanetti et al. Int J Immunopathol Pharmacol. 26(4):889–96, 2013. [92]
Hodgkin Lymphoma	MEM-E/02	Kren L, et al., Pathology, Research and Practice 208: 45–49, 2012. [93]
Chronic Lymphocytic Leukemia	3D12	McWilliams et al., Oncoimmunology. 5(10):e1226720, 2016. [94]
Chronic Lymphocytic Leukemia	3D12	Wagner et al. Cancer, 23(5):814–823, 2017. [52]
Many Cancers	3D12	Sensi M, et al. Int Immunol. 21(3):257–268. 2009. [95]

Table 3.
Expression of HLA-E on human cancer cells (biopsies or cell lines) monitored with commercial mouse anti-HLA-E mAbs (MEM-E/02, MEM-E/06, MEME/07. MEM-E/08, 3D12, 3H2679).

presented in **Figure 4** show that the commercial anti-HLA-E mAbs react with HLA-A, HLA-B and HLA-C in the following order: MEM-06 > MEM-02 > MEM-07 > MEM >08 >> > 3D12. That the mAbs are recognizing the epitopes shared with several HLA-Ia (HLA-A, HLA-B, HLA-C) antigens confirms that none of the above mAbs are specific for HLA-E. Therefore conclusions concerning the expression of HLA-E in human cancers require further validation with monospecific anti-HLA-E mAbs.

5. Anti-HLA-E mAbs: Characteristics, diagnostic and therapeutic potentials

5.1 The technology that clarifies monospecificity or polyreactivity of a mAb of MHC

Luminex multiplex assays are based on xMAP (Multi-Analyte Profiling) technology that enables simultaneous detection and quantitation of antibodies reacting to multiple proteins simultaneously, using detection mAbs [16, 17, 21, 35–39, 96–98]. The results are comparable to assays such as ELISA but with greater specificity, sensitivity and resolution. The technology employs superparamagnetic 6.5-micron microspheres with a magnetic core and polystyrene surface. The beads are

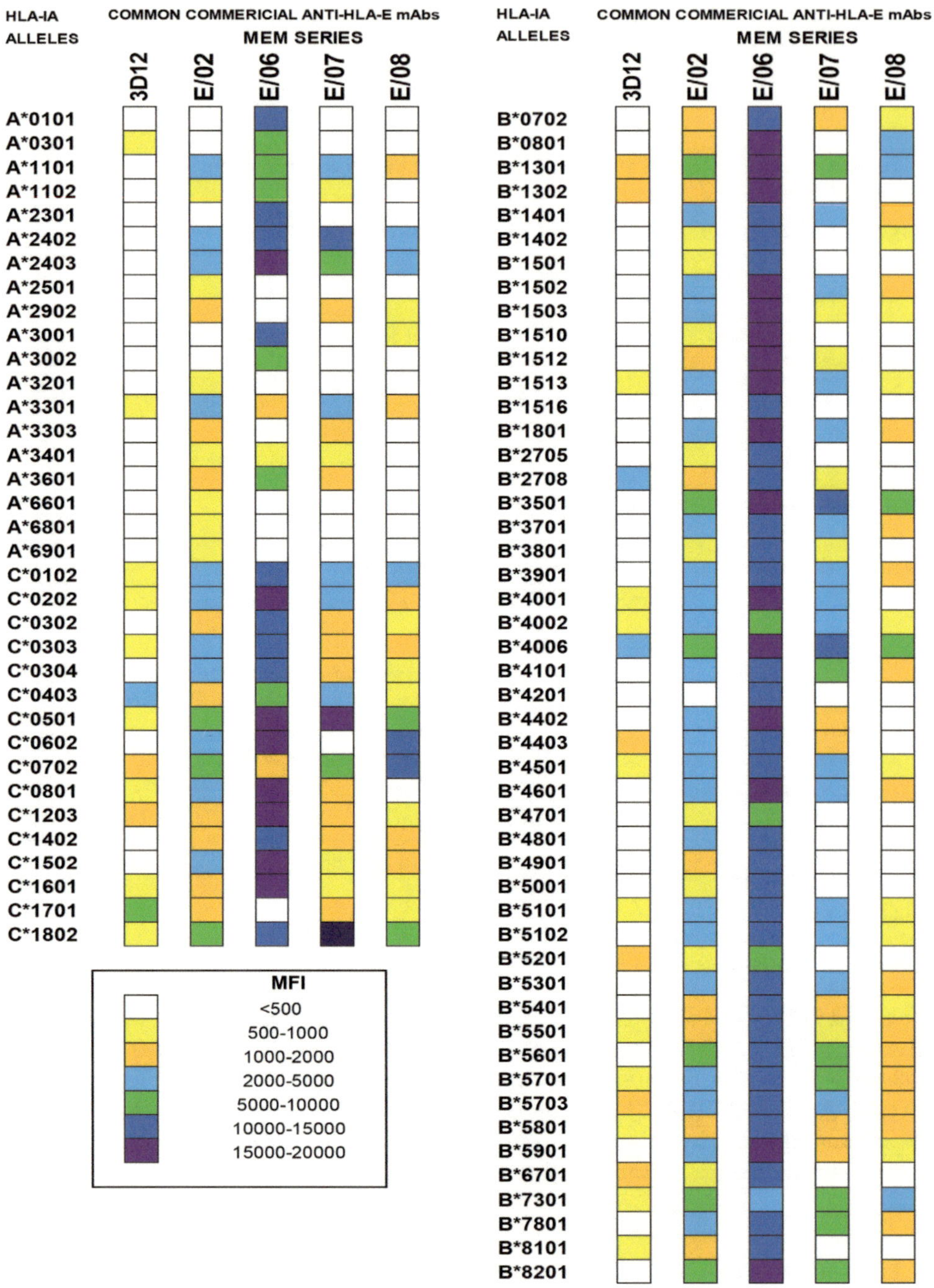

Figure 4.
HLA-IA-polyreactivity of the commercial anti-HLA-E mAbs indicates that these mAbs cannot be considered monospecific or specific for HLA-E. The mAbs were tested at a dilution of 1/300. These mAbs were used to conclude on the expression of HLA-E on human cancers.

internally dyed with precise proportions of red and infrared fluorophores. The Luminex xMAP detection systems identifies differing proportions of the red and infrared fluorophores that result in 100 unique spectral signature microspheres. The antigens are individually attached to polystyrene microspheres by a process of simple chemical coupling. The conjugation of a mAb to one or more of the antigen-coated beads allows it to be evaluated for the mono- or polyreactivity of mAb

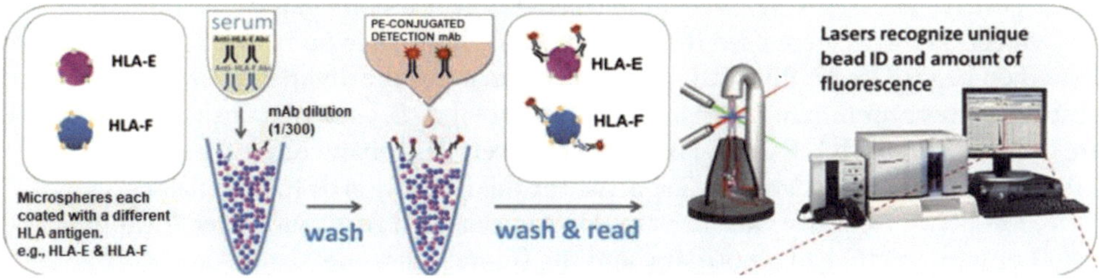

Figure 5.
Luminex single antigen bead assay is used to determine the monospecificity or polyreactivity of the mAbs as well as to determine the strength of the antibodies measured as mean fluorescent intensity (MFI) at specified dilution. The assay is also used to measure the antibody strength titrimetrically. Using peptide inhibition assay epitope affinity or specificity of a mAb can be studied to determine monospecificity or polyreactivity of the mAb. Using a mAb (e.g., HLA-I mAb, TFL-006) recognizing the most commonly shared epitope of an HLA-I (or HLA-II) in an open conformer, the commercial beads can be distinguished as those containing open conformers or closed conformers.

[96–98]. **Figure 5** illustrates the SAB Assay used for determining the monospecificity or polyreactivity of mAbs as well as evaluating the strength of the antibodies measured as mean fluorescent intensity (MFI) at specified dilution. The assay is also used to measure antibody specificity by peptide inhibition assays, to define the epitope-specificity of a mAb. Commercial HLA class I or II beadsets are commercially available as LABScreen (One Lambda Inc., now merged with Thermofisher Inc) and LIFECODES (Immucore Inc)]. The both beadsets together is useful to distinguish CCs from OCs of HLA-I molecules, using a mAb (HLA-I mAb, TFL-006) (See **Table** 7 in [99]).

5.2 Development of mAbs against HLA-E

Following guidelines of the National Research Council's Committee on Methods of Producing Monoclonal Antibodies [35, 98, 100], 235 anti-HLA-E mAbs were generated immunizing mice with recombinant HCs of HLA-E^{R107} (Immune Monitoring Lab, Fred Hutchinson Cancer Research Center, University of Washington, Seattle, WA) (10 mg/ml in MES buffer). In a separate mouse model, HLA-E^{G107} (heavy-chain only) was used as an immunogen. The β2m-free HC of HLA-E (50 μM in 100 mL of PBS (pH 7.4) mixed with 100 mL of TiterMaxVR Gold adjuvant (CytRx, San Diego, CA) were injected into the mouse footpad and intraperitoneum. Three immunizations were given at 12-day intervals. The B cell clones were cultured in RPMI 1640 medium w/L-glutamine and sodium bicarbonate (Sigma-Aldrich, St. Louis, MO, cat. no. R8758), 15% fetal calf serum, 0.29 mg/ml L-glutamine, Pen-Strep (Gemini-Bio, MEd Supply Partners, Atlanta, GA, cat. no. 400–110) and 1 mM sodium pyruvate (Sigma, cat. no. S8636). Several clones were grown using Hybridoma Fusion and Cloning Supplement (HFCS) (Roche Applied Science, Indianapolis, IN, cat. no. 11363735001). The purified-mAbs from HLA-E hybridoma culture supernatants and ascites of hybridoma immunized in BALB/c mice were examined for HLA-I reactivity using Luminex SAB Assay.

5.3 Characterizing the diversity of anti-HLA-E mAbs using single antigen bead (SAB) assay

The HLA-I reactivity of the mAbs was examined by their dose-dependent binding to microbeads coated with 31 HLA-A, 50 HLA-B, and 16 HLA-C antigens and with recombinant single alleles of HLA-E, -F, and -G [35, 98, 100]. The HLA-Ia microbeads have built-in control beads: positive beads coated with human IgG and negative beads coated with serum albumin (human or bovine). For HLA-Ib, the control beads (both positive and negative) were added separately. PE-conjugated anti-human IgG-detection mAbs were used for immunolocalization of mAb bound to HLA antigens coated on beads [35–37, 96–100]. **Table 4** summarizes the diverse types of mAbs observed after immunizing with heavy chains of HLA-E. Group 1 consists of mAbs that are only bound to HLA-E. Anti-HLA-E mAbs were also characterized for their IgG subclasses, using monoclonal IgG specific for the Fc portion of the subclasses

Fluorophore intensity was measured in a specialized flow cytometer (Luminex) together with microbead identifiers, and the fluorescence measurement classified by the bead identifier. Fluorescent intensity generated by Luminex Multiplex Flow Cytometry (LABScan 100) was analyzed using the same computer software and protocols. For each analysis, at least 100 beads were counted. The "trimmed mean" is obtained by trimming a percentage of the high and low ends of distribution and finding the mean of the remaining distribution. Trimmed mean fluorescence intensity (MFI) for the SAB reactions are obtained from output (CSV) file generated by flow analyzer, and it was adjusted for background signal using the formula (sample #N bead – sample negative control bead) [35–37, 96–100]. The MFI was compared with the negative control mean and the standard deviation of MFI recorded. The purpose of MFI is to define the affinity of mAbs to HLAs and the intensity or strength of the mAbs.

5.4 The diversity anti-HLA-E mAbs

Of the 235 hybidomas generated, mAbs secreted by 214 hybridomas were reactive to HLA-E. These mAbs included both monospecific [35, 98] and polyreactive (with other HLA-Ia and HLA-Ib molecules) [98, 101]. **Table 5, A** presents category 1 correspond to monospecific mAbs reacting restrictively to mAbs with HLA-E and failing to recognize HLA-F, HLA-G, HLA-A, HLA-B, and HLA-C. Category 2 refers to HLA-Ib specific anti-HLA-E mAbs (**Table 5, B**). Category 3 presents anti-HLA-E mAbs reactive with several HLA-Ia molecules (HLA-A, HLA-B, and HLA-C) but not reactive to HLA-F and HLA-G (**Table 5, C**). Category 4 presents mAbs recognizing both HLA-Ib and HLA-Ia molecules (**Table 5, D**).

	mAbs formed after immunizing HLA-E						
	HLA Class Ia			HLA Class Ib			
	HLA-A	HLA-B	HLA-C	HLA-E	HLA-F	HLA-G	
Group 1	(−)	(−)	(−)	(+)	(−)	(−)	24 TFL-monospecific anti-HLA-E mAbs
Group 2	(−)	(−)	(−)	(+)	(+)	(−)	TFL-anti-HLA-E/F mAbs
Group 3	(−)	(−)	(−)	(+)	(−)	(+)	TFL-anti-HLA-E/G mAbs
Group 4	(−)	(−)	(−)	(+)	(+)	(+)	TFL-anti-HLA-Ib sepecific mAbs
Group 5	(+)	(+)	(+)	(+)	(−)	(−)	Reactivity of the mAbs 3D12, MEM-E/02 & MEM-E/07 & TFL series
Group 6	(+)	(+)	(+)	(+)	(+)	(−)	Reactivity of the mAb MEM-E/06 & TFL-series
Group 7	(+)	(+)	(+)	(+)	(−)	(+)	Reactivity of the mAb MEM-E/08 & TFL series
Group 8	(+)	(+)	(+)	(+)	(+)	(+)	Reactivity of the mAb TFL-006, TFL-007 & other TFL mAbs

Table 4.
The diverse HLA-E monospecific and polyreactive mAbs generated after immunizing mice with a recombinant heavy chain of HLA-E^{R107} & HLA-E^{G107}.

DOI: http://dx.doi.org/10.5772/intechopen.95235

Nature of mAbs	mAb specificity	number of mAbs	Examples of TFL mAbs	Antigen (heavy chain only) tested on beads	Subclass	HLA-E	HLA-F	HLA-G	HLA-A	HLA-B	HLA-C
						Reactivity in MFI					
A		Antigen immunized: β2-microglobulin-free heavy chain of $HLA\text{-}E^{R107}$									
HLA-E Monospecific mAbs (Category 1)	HLA-E	16	TFL-145, TFL-33, TFL, 34, TFL-73, TFL-74	$HLA\text{-}E^{R}$	IgG1	4 K–22 K	0	0	0	0	0
		3	TFL-001	$HLA\text{-}E^{R}$	IgG2a	0.9 K - 4 K	0	0	0	0	0
			TFL-016								
			TFL-013								
		Antigen immunized: β2-microglobulin-free heavy chain of $HLA\text{-}E^{G107}$									
		5	TFL-185	$HLA\text{-}E^{G}$	IgG1	19 K	0	0	0	0	0
			TFL-184								
			TFL-186								
			TFL-226								
			TFL-254								
B		Antigen immunized: β2-microglobulin-free heavy chain of $HLA\text{-}E^{R}107$									
HLA-IB polyreactive and HLA-IA and non-reactive HLA-E mAbs (Category 2)	HLA-Ib specific mAbs	1	TFL-050	$HLA\text{-}E^{R}$	IgG2b	4 K	3 K	2 K	0	0	0
		Antigen immunized: β2-microglobulin-free heavy chain of $HLA\text{-}E^{G}107$									
		3	TFL-208, TFL-209, TFL-223,	$HLA\text{-}E^{G}$ $HLA\text{-}E^{R}$	IgG1	21 K	8 K	20 K	0	0	0
		4	TFL-164	$HLA\text{-}E^{G}$	IgG2b	14 K–15 K	8 K–9 K	24 K–25 K	0	0	0
			TFL-165								
			TFL-162								
			TFL-161								
	E + G+	1	TFL-191	$HLA\text{-}E^{G}$	NK	1 K	0	1 K	0	0	0
	E + F+	1	TFL-228	$HLA\text{-}E^{G}$	IgG1	19 K	1 K	0	0	0	0

Nature of mAbs	mAb specificity	number of mAbs	Examples of TFL mAbs	Antigen (heavy chain only) tested on beads	Subclass	HLA-E	HLA-F	HLA-G	HLA-A	HLA-B	HLA-C
						Reactivity in MFI					
C	Antigen immunized: β2-microglobulin-free heavy chain of HLA-E^R107										
HLA-IA Polyreactive HLA-E mAbs (Categroy 3)	**E + B + C+**	31	TFL-059	HLA-E^G HLA-E^R	IgG1 (n = 12) IgG2A (n = 9) IgG2b (n = 9) IgG3 (n = 1)	8 K–20 K	0	0	0	1 K–17 K	1 K–7 K
			TFL-143								
			TFL-158								
			TFL-076								
			TFL-159								
	E + A + B + C+	68	TFL-119	HLA-E^G HLA-E^R	IgG1 (n = 27) IgG2A (n = 23) IgG2b (n = 17) IgG3 (n = 1)	11 k–22 k	0	0	1 K–4 K	1 K–24 K	1 K–13 K
			TFL-142								
			TFL-153								
			TFL-118								
			TFL-133								
			TFL-141								
			TFL-095								
	E^G + B+	3	TFL-173	HLA-E^G	IgG1	12 K	0	0	0	1 K	0
			TFL-174								
			TFL-175								
	E + B+	1	TFL-219	HLA-$E^{G/R}$	IgG1	21	0	0	0	2 K	0
	E^G + A + B + C+	6	TFL-167	HLA-E^G	IgG1	15 K-25	0	0	1 K–9 K	1 K 20 K	1 K–20 K
			TFL-170								
			TFL-169								
			TFL-166								

Nature of mAbs	mAb specificity	number of mAbs	Examples of TFL mAbs	Antigen (heavy chain only) tested on beads	Subclass	HLA-E	HLA-F	HLA-G	HLA-A	HLA-B	HLA-C
						Reactivity in MFI					
			TFL-168								
			TFL-205								
	E + A + B + C+	35	TFL-243	HLA-EG/R	IgG1 (n = 22) IgG2A (n = 6) IgG2b (n = 6) IgG3 (n = 1?)	13 K–26 K	0	0	1 K–9 K	1 K–24 K	1 K–20 K
			TFL-246								
			TFL-244								
			TFL-245								
			TFL-172								
			TFL-171								

Nature of mAbs	Immunogen used	mAb specificity	number of mAbs	Examples of TFL mAbs	Subclass	HLA-E		HLA-F	HLA-G	HLA-A	HLA-B	HLA-C
						Reactivity in MFI						
						E^{R107}	E^{G107}					
D. Category 4. HLA- IA and IB polyreactive anti-HLA-E mAbs. (n = 36)												
HLA = IA Polyreactive HLA-IB mAbs (Category 4)	HLA-E^G	E+/F+/G+	4	TFL-232	IgG3	13–22	21	2 to 10	11 to 21	1 to 13	1 to 20	1 to 20
				TFL-177	IgG1 (n = 3)		0					
				TFL-176								
				TFL-198					0			
		E+/G+	16	TFL-236	IgG1 (n = 14)	18–22 (n = 13)	0	0	18–22 (n = 11)	1 to 9	1 to 25	1 to 24
				TFL-238								
				TFL-256	IgG3	22	27		1			
				TFL-229	IgG2b	30	22		18			
		E+/F+	10	TFL-210	IgG1 (n = 10)	18–21	17–19	5 to 11	0	1 to 15	1 to 20	1 to 22

Nature of mAbs	mAb specificity	number of mAbs	Examples of TFL mAbs	Antigen (heavy chain only) tested on beads	Subclass	HLA-E	HLA-F	HLA-G	HLA-A	HLA-B	HLA-C
						Reactivity in MFI					
				TFL-211							
				TFL-212							
				TFL-235							
	HLA-E^{R}	**E+/F+/G+**	3	TFL-049	IgG2b	15–22	8 To 12	2 to 7	1 to 10	1 to 10	1 to 17
				TFL-006	IgG2a (n = 2)						
				TFL-007							
		E+/G+	2	TFL-103	IgG1 (n = 2)	17,18	0	4	1 to 6	1 to 11	1 to 11
				TFL-104							
		E+/F+	1	TFL-063	IgG2b	22	3	0	2-Jan	1 tp 7	3 to 8

mAbs in Bold are highly polyreactive,

Table 5.
Different categories of mAbs (n = 212) formed after immunizing mice with HLA-E open conformer (β2-microglobulin-free heavy chain) of HLA-E^{R107} or HLA-E^{G107}.

5.5 Unique (private) and common (public) epitopes of HLA-E

The international immunogenetics project (http://www.ebi.ac.uk; or http://www.ebi.ac.uk/ipd/imgt/hla/intro.html) updates HLA genes and sequence alleles yearly. We have compared the entire amino acid sequences of HLA-E (**Figure 6**) with 511 alleles of HLA-A, 846 alleles of HLA-B, 275 alleles of HLA-C, 2 alleles of HLA-F, and 2 alleles of HLA-G sequences(see **Table 1**). Amino acid sequences unique to HLA-E (private epitopes) and common amino acid sequences (public epitopes) can be identified by comparing the amino acid sequences of HLA-E with thousands of HLA-Ia and Ib antigens (**Table 6**). Anti-HLA-E mAbs could bind to HLA-E restricted (monospecific) or HLA-I amino acid sequences. Several HLA-E sequences are shared with HLA-A loci or HLA-C loci or specific alleles such as A*3306 or B*8201. **Table 7** shows HLA-E restricted amino acid sequences found in α1 and α2 helices, which were used for peptide inhibition assays. **Figure 7A** illustrates locations of private and public epitopes. **Figure 7B** shows allele-specific amino acid sequences in α1 & α2 helical groove and **Figure 7C** shows shared peptide amino acid sequences.

Peptide inhibition analyses were performed to confirm the monospecificity of HLA-E mAbs. Various concentrations of HLA-E-restricted peptides (serially diluted from the initial concentration of 100 μL to 100 μL) were added to the mAbs (7 μL). The mAbs were further diluted with 14 μL PBS-BSA (pH 7.0; final dilution 1/1200),

------Leader sequence------																					1	2	3	4	5	6	7	8	9
1	2	3	4	5	6	7	8	9	10	11	12	13	14	15	16	17	18	19	20	21	22	23	24	25	26	27	28	29	30
M	V	D	G	T	L	L	L	L	L	S	E	A	L	A	L	T	Q	T	W	A	G	S	H	S	L	K	Y	F	H
10	11	12	13	14	15	16	17	18	19	20	21	22	23	24	25	26	27	28	29	30	31	32	33	34	35	36	37	38	39
31	32	33	34	35	36	37	38	39	40	41	42	43	44	45	46	47	48	49	50	51	52	53	54	55	56	57	58	59	60
T	S	V	S	R	P	G	R	G	E	P	R	F	I	S	V	G	Y	V	D	D	T	Q	F	V	R	F	D	N	D
40	41	42	43	44	45	46	47	48	49	50	51	52	53	54	55	56	57	58	59	60	61	62	63	64	65	66	67	68	69
61	62	63	64	65	66	67	68	69	70	71	72	73	74	75	76	77	78	79	80	81	82	83	84	85	86	87	88	89	90
A	A	S	P	R	M	V	P	R	A	P	W	M	E	Q	E	G	S	E	Y	W	D	R	E	T	**R**	**S**	**A**	**R**	**D**
70	71	72	73	74	75	76	77	78	79	80	81	82	83	84	85	86	87	88	89	90	91	92	93	94	95	96	97	98	99
91	92	93	94	95	96	97	98	99	100	101	102	103	104	105	106	107	108	109	110	111	112	113	114	115	116	117	118	119	120
T	**A**	Q	I	F	R	V	N	L	R	T	L	R	G	Y	Y	N	Q	S	E	A	G	S	H	T	L	Q	W	M	H
100	101	102	103	104	105	106	107	108	109	110	111	112	113	114	115	116	117	118	119	120	121	122	123	124	125	126	127	128	129
121	122	123	124	125	126	127	128	129	130	131	132	133	134	135	136	137	138	139	140	141	142	143	144	145	146	147	148	149	150
G	C	E	L	G	P	D	R	R	F	L	R	G	Y	E	Q	F	A	Y	D	G	K	D	Y	L	T	L	N	E	D
130	131	132	133	134	135	136	137	138	139	140	141	142	143	144	145	146	147	148	149	150	151	152	153	154	155	156	157	158	159
151	152	153	154	155	156	157	158	159	160	161	162	163	164	165	166	167	168	169	170	171	172	173	174	175	176	177	178	179	180
L	R	S	W	T	A	V	D	T	A	A	Q	I	**S**	**E**	**Q**	**K**	**S**	**N**	**D**	**A**	**S**	**E**	A	E	H	Q	R	A	Y
160	161	162	163	164	165	166	167	168	169	170	171	172	173	174	175	176	177	178	179	180	181	182	183	184	185	186	187	188	189
181	182	183	184	185	186	187	188	189	190	191	192	193	194	195	196	197	198	199	200	201	202	203	204	205	206	207	208	209	210
L	E	D	T	C	V	E	W	L	H	K	Y	L	E	K	G	K	E	T	L	L	H	L	E	P	P	K	T	H	V
190	191	192	193	194	195	196	197	198	199	200	201	202	203	204	205	206	207	208	209	210	211	212	213	214	215	216	217	218	219
211	212	213	214	215	216	217	218	219	220	221	222	223	224	225	226	227	228	229	230	231	232	233	234	235	236	237	238	239	240
T	H	H	P	I	S	D	H	E	A	T	L	R	C	W	A	L	G	F	Y	P	A	E	I	T	L	T	W	Q	Q
220	221	222	223	224	225	226	227	228	229	230	231	232	233	234	235	236	237	238	239	240	241	242	243	244	245	246	247	248	249
241	242	243	244	245	246	247	248	249	250	251	252	253	254	255	256	257	258	259	260	261	262	263	264	265	266	267	268	269	270
D	G	E	G	H	T	Q	D	T	E	L	V	E	T	R	P	A	G	D	G	T	F	Q	K	W	A	A	V	V	V
250	251	252	253	254	255	256	257	258	259	260	261	262	263	264	265	266	267	268	269	270	271	272	273	274	275	276	277	278	279
271	272	273	274	275	276	277	278	279	280	281	282	283	284	285	286	287	288	289	290	291	292	293	294	295	296	297	298	299	300
P	S	G	E	E	Q	R	Y	T	C	H	V	Q	H	E	G	L	P	E	P	V	T	L	R	W	K	P	A	S	Q
280	281	282	283	284	285	286	287	288	289	290	291	292	293	294	295	296	297	298	299	300	301	302	303	304	305	306	307	308	309
301	302	303	304	305	306	307	308	309	310	311	312	313	314	315	316	317	318	319	320	321	322	323	324	325	326	327	328	329	330
P	T	I	P	I	V	G	I	I	A	G	L	V	L	L	G	S	V	V	S	G	A	V	V	A	A	V	I	W	R
310	311	312	313	314	315	316	317	318	319	320	321	322	323	324	325	326	327	328	329	330	331	332	333	334	335	336	337		
331	332	333	334	335	336	337	338	339	340	341	342	343	344	345	346	347	348	349	350	351	352	353	354	355	356	357	358		
K	K	S	S	G	G	K	G	G	S	Y	S	K	A	E	W	S	D	S	A	Q	Q	S	E	S	H	S	L		

GenBank: ARB0844 Norman PJ, Norberg SJ, Guethlein LA, Nemat-Gorgani N, Ro	DBSOURCE	accession KY4973ξ Submitted (20-JAN-2017)
FIRST ROW OF NUMBERS BASED ON SECRETED HEAVY CHAINS, SECOND ROW NUMBERS BASED ON GENE SEQUENCE		

Figure 6.
Amino acid sequence of HLA-E^{R107}. Two sets of serial numbers provide one to include leader sequence and another after deleting leader sequence. Sequences in the boxes refer to either specific (private) or shared (public) epitopes. The box with bold letters was used to test for peptide inhibition in our experiments using TFL-monospecific mAbs.

Comparison of the amino acid sequences of HLA-E with other HLA-I antigens							
		HLA alleles					
HLA-E peptide sequences	Number of amino acids	Classical HLA-Ia			Non-classical HLA-Ib		Specificity
		A	B	Cw	F	G	
^{47}PRAPWMEQE^{55}	9	1	0	0	0	0	A*3306 restricted
^{59}EYWDRETR^{65}	8	5	0	0	0	0	A-restricted
^{65}RSARDTA^{71}	6	**0**	**0**	**0**	**0**	**0**	**E-monospecific**
^{90}AGSHTLQW^{97}	8	1	10	48	0	0	Multispecific
^{108}RFLRGYE^{123}	7	24	0	0	0	0	A-restricted
^{115}QFAYDGKDY^{123}	9	1	104	75	0	0	Multispecific
^{117}AYDGKDY^{123}	7	491	831	271	21	30	Highly Multispecific
^{126}LNEDLRSWTA^{135}	10	239	219	261	21	30	Multispecific
^{137}DTAAQI^{142}	6	0	824	248	0	30	Multispecific
^{137}DTAAQIS^{143}	7	0	52	4	0	30	Multispecific
^{143}SEQKSNDASE^{152}	10	**0**	**0**	**0**	**0**	**0**	**E-monospecific**
^{157}RAYLED^{162}	6	0	1	0	0	0	B*8201-restricted
^{163}TCVEWL^{168}	6	282	206	200	0	30	Multispecific
^{182}EPPKTHVT^{190}	8	0	0	19	0	0	C-restricted

Table 6.
Identifying HLA-E specific epitope or amino acid sequences: Peptide sequences specific and shared between HLA-E and HLA class Ia alleles: Monospecific (HLA-E restricted) versus polyreactive epitopes.

and then exposed to 2 mL of beads. The two different HLA-E-restricted peptides, RSARDTA and SEQKSNDASE were synthesized and purified by GenScript Corporation (Piscataway, NJ). The assay was performed in triplicate. Dosimetric peptide inhibition analysis was performed for mAb TFL-033. Before dosimetric peptide inhibition, the mAb TFL-033 was dosimetrically titrated to assess their strength (MFI), and protein-G purified culture supernatants and ascites compared. Then, concentrated Protein-G purified from ascites is titrated and the protein content is measured. Titrimetric inhibition was done with ascites protein-G concentrate. A summary of the peptide inhibition experiments is presented in **Figure 8**. Results confirm that TFL-003 binding to HLA-E can be inhibited dosimetrically using two HLA-E-restricted epitopes. The level of inhibition differed between the two epitopes.

5.6 Diagnostic potential of HLA-E monospecific mAbs

Immunolocalization of HLA-E on human melanoma cancer tissues was performed using culture supernatants (s) or ascites (a) of TFL monospecific mAbs (TFL-033, TFL-034, TFL-074, and TFL-216), and staining is compared with commercial anti-HLA-E mAb (MEM-E/02) [35, 98]. Titration of Protein-G purified culture supernatants and ascites concentrates of different anti-HLA-E monospecific mAbs are shown in **Table 8**. As revealed in **Figure 4**, the MEM-02 cross-reacts with several HLA class Ia alleles. Although it stains melanoma tissues, due to the paucity of HLA-E specificity, specific localization of HLA-E was confirmed with monospecific anti-HLA-E mAbs (**Figure 9A**). Similarly, immune-localization of HLA-E on human

	Peptide [# 1] specific for HLA-E						Peptide [# 2] specific for HLA-E									
HLA Class Ib	α1	α1	α1	α1	α1	α1	**HLA Class Ib**	α2	α2	α2	α2	α2	α2	α2	α2	α2
	65	66	67	68	69	70		143	144	145	146	147	148	149	150	151
E*01010101	*R*	*S*	*A*	*R*	*D*	*T*	**E*01010101**	*S*	*E*	*Q*	*K*	*S*	*N*	*D*	*A*	*S*
G*01010101	*R*	N	T	K	A	H	**G*01010101**	**S**	K	R	**K**	C	E	A	**A**	N
F*01010101	G	Y	A	K	A	N	**F*01010101**	T	Q	R	F	Y	E	A	E	E
A*110101	**R**	N	V	K	**A**	Q	**A*110101**	T	K	R	**K**	W	E	A	**A**	H
B*1401	Q	I	C	K	T	N	**B*1401**	T	Q	R	**K**	W	E	A	**A**	R
B*350101	Q	I	F	K	T	N	**B*350101**	T	Q	R	**K**	W	E	A	**A**	R
B*40060101	Q	I	S	K	T	N	**B*40060101**	T	Q	R	**K**	W	E	A	**A**	R
B*530101	Q	I	F	K	T	N	**B*530101**	T	Q	R	**K**	W	E	A	**A**	R
B*5801	R	N	M	K	**A**	S	**B*5801**	T	Q	R	**K**	W	E	A	**A**	R
CW*050101	Q	K	Y	K	R	Q	**CW*050101**	T	Q	R	**K**	W	E	A	**A**	R
CW*080101	Q	K	Y	K	R	Q	**CW*080101**	T	Q	R	**K**	W	E	A	**A**	R
CW*1802	Q	K	Y	K	R	Q	**CW*1802**	T	Q	R	**K**	W	E	A	**A**	R
Qa-1(murine eq:HLA-E)	W	K	**A**	**R**	D	M	**Qa-1(murine eq:HLA-E)**	**S**	K	H	**K**	S	E	A	V	D

Table 7.
Identifying HLA-E specific epitope or amino acid sequences: Comparing the two HLA-E restricted sequences with other HLA-I amino acid sequences at the same position.

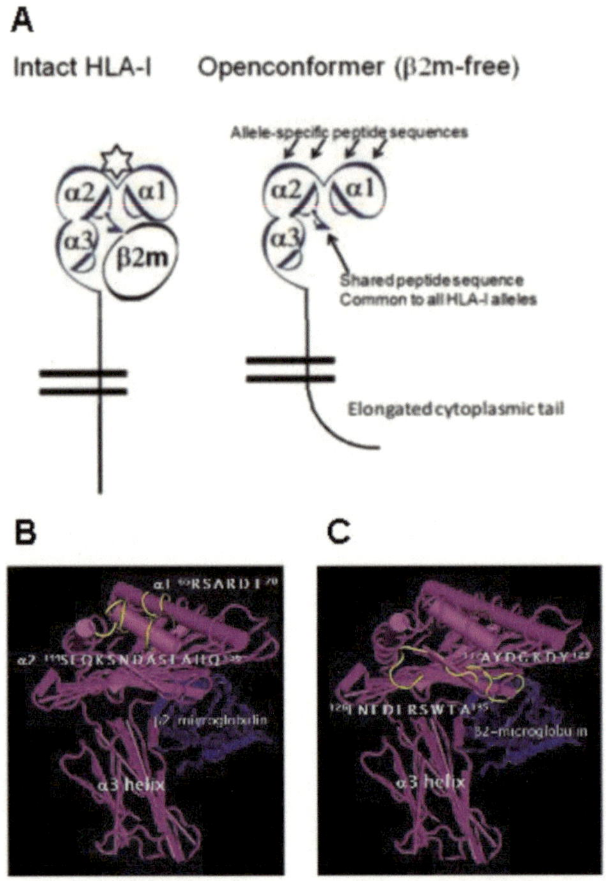

Figure 7.
Diagrammatic illustration of the structure of HLA-E, closed (intact trimer) and open conformers and specific (private) and shared (public) epitopes. (A) Illustrates the locations of allele-specific sequence (private epitope) and shared peptide (public epitopes) sequence. HLA-E with β2-microglobulin (in blue) showing (B) the allele-specific amino acid sequences (private epitopes) in α1 & α2 helical groove and (C) shared peptide amino acid sequences (public epitopes).

gastric diffused carcinoma paraffin tissue sections was observed after staining with the diluted ascites of monospecific mAb TFL-033a and MEM-E/02. The reliability of HLA-E tissue localization with monospecific immunostaining of human gastric adenocarcinoma (A, B) with TFL-033 and MEM-E/02 with that obtained for gastric diffuse carcinoma (C, D) control, stained without primary mAbs. MEM-E/02 failed to stain any cells while TFL-033a showed intense and widely distributed staining indicating the overexpression of intact HLA-E (**Figure 9C**). Immunostaining was performed on human breast ductal adenocarcinoma with TFL monospecific-mAbs and results obtained using monospecific anti-HLA-E mAb TFL-216, generated by immunizing HLA-E^G, is presented in **Figure 9D**.

Detailed immunodiagnostic analyses were performed using a tissue microarray (TMA) of normal gastric mucosal and primary gastric cancer tissues [98]. Three tissue microarrays (TMAs; US Biomax, Rockville, MD) were carefully selected. The tissue sections of all TMA were 1.5 mm in diameter and 5 μm thick. In TMA of normal gastric mucosa and of primary gastric cancer, which contained 30 adenocarcinomas, 40 diffuse carcinomas and ten normal gastric mucosae were immunostained. TMA array included: well-differentiated, moderately differentiated, poorly differentiated, and undifferentiated cancer. In addition, TMA also

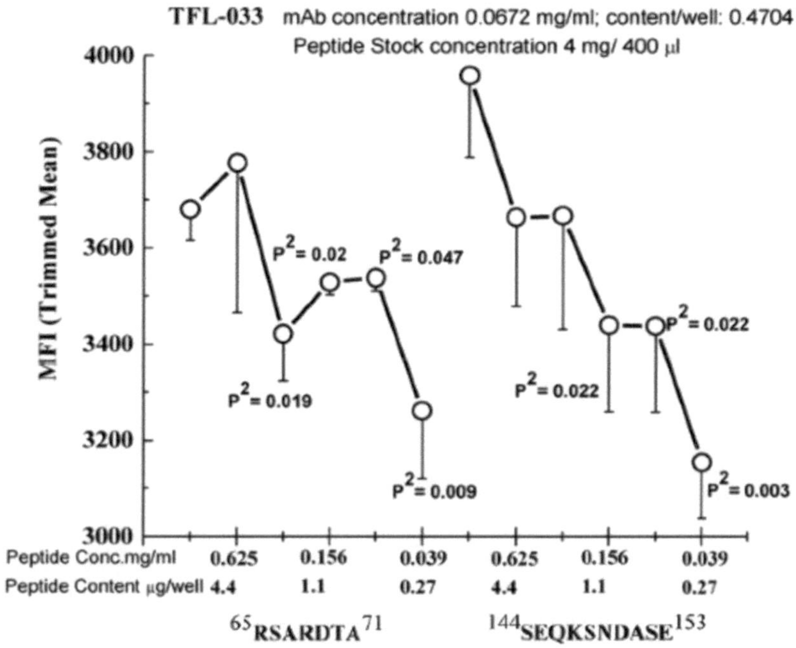

Figure 8.
Dosimetric inhibition of purified culture supernatants of TFL-033 with two HLA-E-restricted peptides, ^{65}RSARDTA^{71} and ^{143}SEQKSNDASE^{152}, at concentrations ranging from 4.4 to 0.27 mg/well. Although both peptides showed inhibition, the α2 helical peptide SEQKSNDASE showed better dosimetric inhibition than the other peptide. Peptide concentration and peptide content (μG/well) in parenthesis are shown. Pair-sample or equal-variant t-tests were carried out in this investigation using a graphic website (www.originlab.com). (*Source*: U. S. Patent No 10,656,158 B2 (*U.S. patent application No. 13/507,537)* issued on May 19, 2020, to Dr. Mepur H. Ravindranath) *see also Int J cancer. 2014;134(7):1558–70. DOI: 10.1002/ijc.28484.*

included Stages I to IV of metastatic gastric cancer with 5 peritoneal, 3 liver, 27 lymph node metastases. TMA was immunostained with TFL-033 mAbs (culture supernatants and ascites), controls were stained without primary mAbs [98]. The diagnostic potential of HLA-E-monospecific mAb TFL-033 for different kinds and stages of gastric cancer is illustrated in **Figure 4a** in International Journal of Cancer [98]. The observations confirm that specific identification and localization of MHC antigens, stringently require monospecific mAbs. The conclusion is highly reliable compared to the use of polyreactive commercial mAbs (MEM-E/02) [36, 98], presented in **Figure 4**. Importantly, characterizations of monospecificity should include (1) multiantigen coated solid matrix assays, e.g., Luminex multiplex SAB assay; (2) titrimetric inhibition with the private epitope of the antigen. Only such monospecific mAbs are reliable for diagnosis and therapeutic purposes.

5.7 Differences in the immunoregulatory potentials of HLA-E monospecific versus polyreactive mAbs

5.7.1 Potential of polyclonal anti-HLA-E mAbs in immune regulation

Immunoregulatory properties of both monospecific (TFL-033) and polyreactive (TFL-006 & TFL-007) anti-HLA-E mAbs were examined for their ability to suppress or activate CD3/CD4+, CD3/CD8+ T cells, T-regs, and CD3+/CD19/20+ B cells. The results show that the polyreactive anti-HLA-E mAbs (TFL-006/TFL-007) are immunosuppressive comparable to IVIg, used in immunotherapy of several diseases [16, 17]. Indeed the anti-HLA antibody profile of IVIg from different sources showed

Sample	Dilution	TFL-033	TFL-034	TFL-073	TFL-074
Culture Supernatant	*Neat*	11273	11601	7781	8493
Protein-G purified Culture supernatant	*(1:10)*	4424	2730	1974	2507
Protein-G purified Culture supernatant Concentrate	*(1:10)*	11953	10364	7708	8467
	(1:20)	9423	8146	6861	7500
	(1:40)	8167	6347	5324	5883
	(1:80)	6203	4622	3792	4176
	(1:160)	4139	1379	2683	2438
	(1:320)	2862	626	1454	943
	(1:640)	1434	198	590	474
	(1:1280)	694	98	275	220
Protein-G purified Ascites Concentrate (Eluate # 2)	*(1:50)*	17898			
	(1:100)	16246			
	(1:200)	14004			
	(1:400)	12520			

Table 8.
Titration of protein-G purified culture supernatant and ascites concentrates of different HLA-E monospecific mAbs. These concentrates were used for immunolocalization, peptide inhibition studies as well as for their effects on T-lymphoblasts.

both HLA-Ia and HLA-Ib reactivities [16, 17]. IVIg preparations were reported to suppress CD4+ T cells [102–113], CD20+ B cells [108–113] and expand CD4 + CD25+ T-regs [114, 115]. The polyreactive anti-HLA-E mAbs performed the major immunoregulatory functions better than IVIg [101, 116–118]. These functions are (1) suppression of CD19+ B lymphocyte blastogenesis, proliferation, and suppression of production of anti-HLA-I and anti-HLA-II IgG Abs, (2) suppression of blastogenesis and proliferation of CD4+ as well as CD8+ T lymphocytes, and (3) expansion of CD4 +. CD25+ and FoxP3+ T-regs. The monospecific mAbs, when used as controls failed to perform these functions. Peptide inhibition analyses revealed that mAbs TFL-006 and TFL-007 bind to shared amino acid sequences of HLA-I molecules (^{117}AYDGKDYLT^{125}, ^{126}LNEDLRSWTAV^{136}, and ^{137}DTAAQI^{142}) (**Figure 7C**). Possibly such binding affinity of polyreactive but not monospecific mAbs contributes to the unique immunoregulatory functions mimicking IVIg [101, 118].

5.7.2 Therapeutic potential of anti-HLA-E monospecific mAbs

In contrast to polyreactive anti-HLA-E mAb, monospecific mAbs (TFL-033) recognized HLA-E- specific amino acid sequences (^{65}RSARDT^{70} and^{154}AESADNSKQES^{144}) on the α1 and α2 helices (**Figure 7B**).

5.7.2.1 Monospecific mAbs promote the proliferation of CD8+ T lymphocytes

To test whether monospecific anti-HLA-E mAbs suppress proliferation of the CD3+, CD4+, or CD8+ T cells, human T lymphocytes (both CD4+ and CD8+) isolated from whole blood of a normal male donor with Ficol Hypaque (31) were treated either with phytohaemagglutinin (PHA, EY Laboratories, San Mateo, CA) at a final concentration of 2.25 mL/mL or not exposed to PHA (31). The mAbs (monospecific mAbs TFL-033, TFL-034, TFL-073, TFL-074, and TFL-216, polyreactive mAb TFL007, and negative control antibodies) were separately added to cells in culture within 2 hours after adding

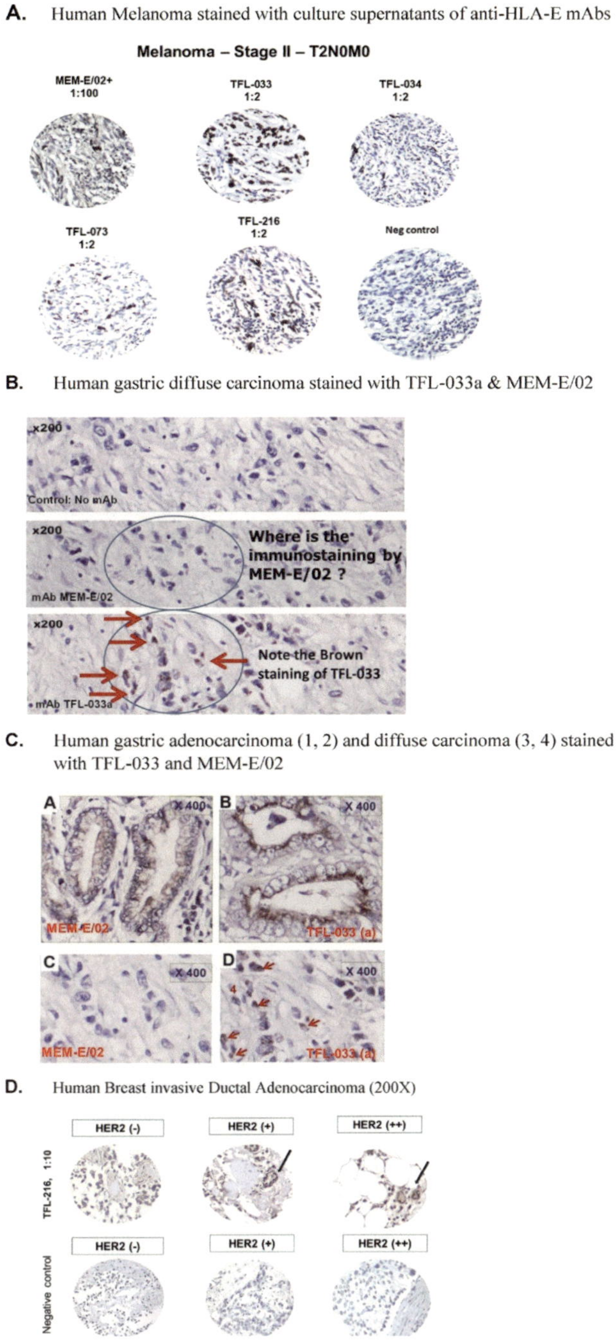

Figure 9.
*Immunolocalization of HLA-E in cancer tissues with culture supernatants (s) or ascites (A) of TFL monospecific mAbs compared with staining by MEM-E/02, an HLA-E mAb that shows cross-reactivity to HLA class Ia alleles. (**A**) Human melanoma paraffin tissue sections stained with the culture supernatants of TFL monospecific MAbs and MEM-E/02. (**B**) Human gastric cancer (diffused carcinoma) paraffin tissue sections stained with the diluted ascites of monospecific MAb TFL-033a and MEM-E/02. (**C**). Immunostaining of human gastric adenocarcinoma (A, B) and gastric diffuse carcinoma (**C, D**) control, stained without primary mAbs. Note the differences in staining between the two antibodies; MEM-E/02 failed to stain any cells while TFL-033a showed intense and widely distributed staining indicative of overexpression of intact HLA-E. (**D**) Human breast ductal adenocarcinoma stained with monospecific anti-HLA-E mAb TFL-216 generated by immunizing HLA-E^{G}. (source:* U. S. Patent No 10,656,158 B2 *(U.S. patent application No. 13/507,537)* issued on May 19, 2020, to Dr. Mepur H. Ravindranath*) see also Int J cancer. 2014;134(7):1558–70. DOI: 10.1002/ijc.28484.*

Presence or absence of CD4/CD8	**CD3+ NAÏVE T-CELLS**				**CD3+ LYMPHOBLASTS**							
	No PHA		**With PHA**		**No PHA**				**With PHA**			
	CD4+/ CD8-	**CD4-/ CD8+**	**CD4+/ CD8-**	**CD4-/ CD8+**	**CD4+/ CD8-**	**CD4-/ CD8+**	**CD4+/ CD8+**	**CD4-/ CD8-**	**CD4+/ CD8-**	***CD4-/ CD8+***	**CD4+/ CD8+**	**CD4-/ CD8-**
No mAb *[n = 5]*												
Mean	3063	547	1249	475	197	*65*	141	52	867	*325*	128	289
SD	149	86	99	37	33	*14*	35	15	115	*126*	43	84
2-tail p [<]		*<0.0001*							***0.001***	*0*	*NS*	*0*
mAb TFL-033 (IgG1) [n = 3]												
[1/30]												
Mean	3185	755	1170	536	223	***163***	153	99	1129	***505***	152	412
SD	180	146	58	12	40	*27*	80	13	86	*23*	16	20
2-tail p [<]	*NS*	*NS*	*NS*	***0.009***	*NS*	***0.015***	*NS*	*0.005*	***0.010***	***0.016***	*NS*	*0.014*
[1/150]												
Mean	3238	681	1149	508	252	***120***	205	68	1266	***572***	157	412
SD	14	64	21	22	30	*17*	13	9	80	*31*	14	16
2-tail p [<]	*NS*	*NS*	*NS*	***NS***	*0.047*	***0.001***	*0.020*	*NS*	***0.001***	***0.003***	*NS*	*0.001*
mAb TFL-007 (Polyreactivec anti-HLA-E, IgG2a) [n = 3]												
[1/10]												
Mean	2876	451	1183	444	164	*63*	145	52	676	*317*	100	222
SD	136	72	19	26	33	*2*	3	17	79	*25*	4	29
2-tail p [<]	*NS*	*NS*	*NS*	*NS*	*NS*	*NS*	*NS*	*NS*	*0.027*	*NS*	*NS*	*NS*
[1/50]												

Presence or absence of CD4/CD8	CD3+ NAÏVE T-CELLS					CD3+ LYMPHOBLASTS						
	No PHA		With PHA			No PHA			With PHA			
	CD4+/ CD8-	CD4-/ CD8+	CD4+/ CD8-	CD4-/ CD8+	CD4+/ CD8-	CD4-/ CD8+	CD4+/ CD8+	CD4-/ CD8-	CD4+/ CD8-	CD4-/ CD8+	CD4+/ CD8+	CD4-/ CD8-
Mean	3,088	667	1,075	491	230	*107*	193	80	892	*443*	122	339
SD	65	16	55	48	23	7	17	4	26	*18*	8	21
2-tail p [<]	*NS*	*0.018*	*0.013*	*NS*	*NS*	*NS*	*0.019*	*0.006*	*NS*	*NS*	*NS*	*NS*

Table 9.
TFL-033 promotes T-lymphoblast proliferation of CD8+ naïve T cells and T-Lymphoblasts in the absence or the presence of PHA. The proliferation of CD4+ T lymphoblasts occurs only after PHA activation.

PHA (final 200 mL) (31). Detailed experimental protocol is described elsewhere (31). The effects of mAbs (monospecific mAb TFL-033 and polyreactive mAb TFL-007) on untreated (no PHA) and PHA-treated T lymphocytes in these categories of T cells: CD4+/CD8-, CD4-/CD8 +, CD4 + /CD8 +, and CD4-/CD8- are presented in **Table 9**. There was a significant increase in numbers of CD4-/CD8+ T lymphoblasts among the PHA-treated T lymphoblasts under the influence of TFL-033 s at 1:30 and 1:150). Numbers of PHA-untreated T lymphoblasts increased for almost all mAbs, TFL-033 s at 1/30 and 1/150, TFL-034 s at 1/10 and 1/50, TFL-073 s at 1/50, TFL-074 s at 1/10 [35]. An increase in PHA-untreated T lymphoblasts clarifies the functional potential of HLA-E monospecific mAbs in augmenting CD4−/ CD8+ T lymphoblasts. A significant increase in numbers of PHA-treated CD3+/CD4-/CD8+ lymphoblasts suggests that monospecific monoclonal mAbs, particularly TFL-003 confers the potential to augment cytotoxic T cells. Results prompt investigating humanized version TFL-003 on proliferation cytotoxic T-cells.

5.7.2.2 HLA-E expressed on cancer cells can directly bind to CD8+ T cells and NK cells and suppress their tumor-killing activity

Cancer cells lose their cell surface HLA-Ia alleles (HLA-A, HLA-B, and HLA-C) and upregulate the surface expression of HLA-Ib molecules (HLAE, HLA-F, and HLA-G) [57, 82, 119–128]. The upregulation of HLA-E gene expression is correlated with immunolocalization and overexpression of cell surface HLA-E [71, 91, 128–132]. HLA-E gene expression in some cancers [e.g., melanoma] is ranked 19th among overexpressed genes [133]. HLA-E overexpression and loss of HLA-Ia in

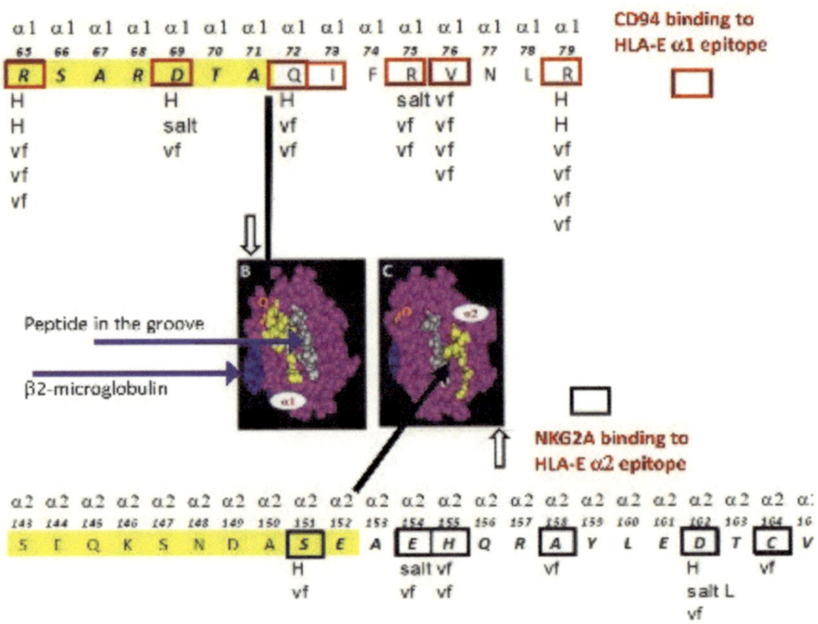

Figure 10.
Binding of HLA-E to the inhibitory receptors CD94 and NKG2A on both CD8+ CTLs and NKT cells. The structural configuration of the binding of HLA-E and the inhibitory receptors, leading to the arrest of the anti-tumor activity function of CD8+ and NKT cells. The interaction between HLA-E and the inhibitory receptors involves the binding of amino acids located on the α1 and α2 helices of HLA-E to specific amino acids on CD94 and NKG2A. The amino acid sequences on HLA-E recognized by the inhibitory receptors are unique and specific for HLA-E and they are also recognized by HLA-E monospecific mAbs. The binding involves H-bonding (H), van der Waal forces (vf), and salt linkages (salt) of the amino acids of HLA-E a1 and a2 helices and CD94 and NKG2A inhibitory receptors. (Modified from Ravindranath et al. Monoclon Antib Immunodiagn Immunother. 2015,34(3):135–53).

cancer cells are correlated with disease progression and poor prognosis [60, 82, 130, 134]. Disease progression is attributed to the suppression of the tumor-killing activity of CD8+ cytotoxic T lymphocytes (CTLs) and NKT cells.

Cell surface and soluble HLA-E are capable of binding to the inhibitory receptors CD94 and NKG2A on both CTLs (CD3+/CD8+), NK cells (CD2+, CD7+, CD11b+, CD11c+, CD90+, perforin+, & granzyme A+) and NKT cells (plus CD8+) [25, 27, 135, 136]. These cells are capable of destroying tumor cells. These cells interact with MHC-I ligands (HLA-E) on tumor cells through inhibitory receptors. The binding of above mentioned immune cells to HLA-E overexpressed on tumor cells cell surface may explain why the cancer patients failed to respond to NK cell therapies.

Interaction between HLA-E and inhibitory receptors involves the binding of HLA-E specific amino acids located on α1 and α2 helices (**Table 7**) to specific amino acids on CD94 and NKG2A (**Figure 10**) [22, 27, 135, 136]. This specific interaction is attributed to the loss of anti-tumor activity of CD8+ CTLs as well as that of NK or NKT cells [22, 27, 135, 136]. We have used the synthetic peptides of these sequences to ascertain the specific binding affinity of anti-HLA-E mAbs (**Figure 8**). The ability of monospecific anti-HLA-E mAbs to bind at the site of epitopes of CD94 and NKG2A on HLA-E favor the use of the monospecific anti-HLA-E mAbs to mask binding sites of inhibitory receptors on HLA-E. Such blocking of HLA-E may help restore the antitumor efficacy of NK cells and CD8+ T cells that were lost due to the interaction of inhibitory receptors and HLA-E. Possibly humanized monospecific anti-HLA-E may be potentially considered for anti-cancer NK therapy.

6. Conclusion

The anti-HLA-E mAbs TFL- 033, TFL-034, TFL-073, and TFL-074 due to their monospecificity are advantageous than the commercial anti-HLA-E mAbs for specific identification and localization of HLA-E on the surface of human cells, particularly in different cancer types. Our observations stress the need for characterization of monospecificity and epitope specificity of any mAb, after analyzing binding affinity on a multiplex solid matrix assays coated with the desired antigen (in question) and the closely related antigens and inhibition of the binding affinity using peptides sequences specific for the antigen in question. This is an important criterion to be followed for all clinical diagnostic and therapeutic antibodies. If specific epitopes are exposed to antigen located on the cell surface, it would be a more valuable diagnostic tool, than those binding to specific but cryptic epitopes.

The HLA-E monospecific antibodies (e.g., TFL-033) are capable of augmenting proliferation of non-activated CD8+ T cells and activated CD8+ T-lymphoblasts. TFL-033 binds to a unique epitope of HLA-E, a region that is involved in binding to inhibitory receptors (CD94 and NKG2A) present on CD3+/CD8+ T cells (Cytotoxic T cells) and CD3-/CD8+ NKT cells and NK cells. The binding of HLA-E to inhibitory receptors results in the suppression of anti-tumor cytotoxic functions of these immune cells. *Since TFL-033 can also upregulate anti-tumor cytotoxic T cell lymphoblasts and also capable of blocking the interaction between cancer-associated HLA-E and inhibitory receptors CD94/NKG2A, the mAb can be considered as a double-edged sword to eliminate cancer cells.* Therefore, TFL-033 could be a valuable therapeutic agent for passive immunotherapy of human cancer, provided the mAb is humanized.

In contrast to monospecific mAbs, HLA-I polyreactive anti-HLA-E monoclonal Abs (TFL-006 and TFL-007) mimic not only HLA-I reactivity of IVIg but also performs several critical immunoregulatory functions of IVIg, better than IVIg *per se*. These functions include suppression of blastogenesis and proliferation of CD4+ T cells and CD8+ T cells, effective inhibition of production of anti-HLA-I and

HLA-II Abs. HLA-I polyreactive anti-HLA-E monoclonal Abs (TFL-006 and TFL-007) are capable of upregulating T-regs. T-regs acting alone is capable of suppressing CD4+ T cells, CD8+ T cells, and antibody.

Acknowledgements

We wish to express sincere thanks to Professor Dr. Edwin L. Cooper, *Distinguished Professor, at* Collaborative Centers for Integrative Medicine, University of California, Los Angeles, for critically editing this manuscript. We wish to dedicate this chapter to Professor Edwin L Cooper. We also wish to thank Professor Mark Terasaki and Terasaki Family Foundation for providing financial support to carry out the research reported in this chapter at Terasaki Foundation Laboratory (TFL), Santa Monica, CA. We wish to thank the following research scientists at TFL, Drs. Dong Zhu, Toshiyuki Sasaki, Vadim Jucaud, Mr. Tho Pham, Mr. Satoru Kawakita, Mr. Curtis Y. Maehara, and Mrs. Judy Hopfield were directly involved in the experiments reported in this chapter.

A patent was filed based on our research on monospecific anti-HLA-E mAb TFL-033 and a U. S. Patent No 10,656,158 B2 was issued on May 19, 2020, to Dr. Mepur H. Ravindranath & Late Professor Paul Ichiro Terasaki. Hybridoma of TFL-033 is deposited with ATCC Patent Depository (ID: PTA-125908) at Manassas, Virginia 20110, USA.

Conflict of interest

The authors declare no conflict of interest.

Author details

Mepur H. Ravindranath[1*†] and Fatiha E.L. Hilali[2]

1 Children's Hospital, Formerly, Terasaki Foundation Laboratory, Los Angeles, CA

2 28 DB Dighoussi Zitoune Meknes, Morocco

*Address all correspondence to: thiruranganath04@gmail.com

†Formerly, Recipient of Fulbright Award at the University of Madras India and came to the University of California, Los Angeles, and appointed as Adjunct Associate Professor in the Department of Anatomy. Senior Research Scientist (2008-2018), Terasaki Foundation Laboratory, Santa Monica, CA, Director, Department of Glycoimmunomics (1990-2008), John Wayne Cancer Institute, Santa Monica CA. Adjunct Faculty, University of California, Los Angeles (1983-1990), Faculty and Fulbright Recipient (1983-1984), University of Madras, Chennai, Tamilnadu, India 60005 (1972-1983).

References

[1] Marsh SGE, Parham P, Barber LD. The HLA. Facts Book. Academic Press. San Diego, 2000, 400p.

[2] Arosa FA, Esgalhado AJ, Padrão CA, Cardoso EM. Divide, Conquer, and Sense: CD8+CD28- T Cells in Perspective. Front Immunol 2017;7:665, DOI:10.3389/fimmu. 2016.00665.

[3] Arosa FA, Santos SG, Powis SJ. Open conformers: the hidden face of MHC-I molecules. Trends Immunol 2007;28(3): 115–123, DOI:10.1016/j.it. 2007 01.002.

[4] Schnabl E, Stockinger H, Majdic O, Gaugitsch H, Lindley IJ, Maurer D, et al., (1990) Activated human T lymphocytes express MHC class I heavy chains not associated with beta 2microglobulin. J Exp Med, 1990;171(5): 1431–1442. DOI: 10.1084/jem.171.5.1431.

[5] Schumacher TN, Heemels MT, Neefjes JJ, Kast WM, Melief CJ, Ploegh HL. (1990) Direct binding of peptide to empty MHC class I molecules on intact cells and in vitro. Cell 1990;62(3);563–567. DOI:10.1016/0092-8674(90) 90020-F

[6] Benjamin RJ, Madrigal JA, Parham P. Peptide binding to empty HLA-B27 molecules of viable human cells. Nature, 1991;351(6321);74–77. DOI: 10.1038/351074a0.

[7] Majdic O, Schnabl E, Stockinger H, Gadd S, Maurer D, Radaszkiewics T, et al. LA45, an activation-induced human lymphocyte antigen with strong homology to MHC class I molecules. In Leukocyte Typing IV. Oxford University Press, 1989, 511p.

[8] Madrigal JA, Belich MP, Benjamin RJ, Little AM, Hildebrand WH, Mann DL, et al., Molecular definition of a polymorphic antigen (LA45) of free HLA-A and -B heavy chains found on the surfaces of activated B and T cells. J Exp Med 1991;174(5);1085–1095, DOI: 10.1084/jem.174.5.1085.

[9] Raine T, Brown D, Bowness P, Hill Gaston JS, Moffett A, Trowsdale J, et al. Consistent patterns of expression of HLA class I free heavy chains in healthy individuals and raised expression in spondyloarthropathy patients point to physiological and pathological roles. Rheumatology (Oxford) 2006;45(11); 1338–1344, DOI: 10.1093/rheumatology/kel305

[10] Cardoso EM, Esgalhado AJ, Patrao L, Santos M, Neves VP, Martinez J, et al. Distinctive CD8+ T cell and MHC class I signatures in polycythemia vera patients. Ann. Hematol. 2018;97(9): 1563–1575, DOI: 10.1007/s00277-018-3332-7

[11] Santos SG, Powis SJ, Arosa FA. Misfolding of major histocompatibility complex class I molecules in activated T cells allows cis-interactions with receptors and signaling molecules and is associated with tyrosine phosphorylation. J Biol Chem. 2004; 279 (51):53062–53070. DOI: 10.1074/jbc. M408794200.

[12] Goodridge JP, Lee N, Burian A, Pyo CW, Tykodi SS, Warren EH, Yee C, Riddell SC, Geraghty DE. HLA-F and MHC-I open conformers cooperate in an MHC-I antigen cross-presentation pathway. J. Immunol. 2013;191:1567–1577, DOI:10.4049/jimmunol.1300080.

[13] Brynjolfsson SF, Persson Berg L, Olsen Ekerhult T, Rimkute I, Wick MJ, Mårtensson IL, Grimsholm O. Long-Lived Plasma Cells in Mice and Men. Front Immunol. 2018;9:2673. DOI: 10.3389/fimmu.2018.02673.

[14] Morales-Buenrostro LE, Terasaki PI, Marino-Vázquez LA, Lee JH, El-Awar N, Alberú J. "Natural" human leukocyte

antigen antibodies found in nonalloimmunized healthy males. Transplantation. 2008;86(8):1111–1115. doi: 10.1097/TP.0b013e318186d87b.

[15] Ravindranath MH, Kaneku H, El-Awar N, Morales-Buenrostro LE, Terasaki PI. Antibodies to HLA-E in nonalloimmunized males: pattern of HLA-Ia reactivity of anti-HLA-E-positive sera. J Immunol. 2010;85(3): 1935–1948. DOI: 10.4049/jimmunol.1000424.

[16] Ravindranath MH, Terasaki PI, Pham T, Jucaud V, Kawakita S. Therapeutic preparations of IVIg contain naturally occurring anti-HLA-E antibodies that react with HLA-Ia (HLA-A/-B/-C) alleles. Blood. 2013;121 (11);2013-28. DOI: 10.1182/blood-2012.08-447771.

[17] EL Hilali F, Jucaud V, EL Hilali H, Bhuiyan MH, Mancuso A, LiuSullivan N, Elidrissi A, Mazouz H. Characterization of the Anti-HLA Class I and II IgG Antibodies in Moroccan IVIg Using Regular Beads and Ibeads in Luminex Multiplex Single Antigen Immunoassay Int. J. Immunol. 2017; 5(4): 53–65. DOI: 10.11648/j.iji.20170504.11

[18] Geraghty DE, Koller BH. Orr HT. A human major histocompatibility complex class I gene that encodes a protein with a shortened cytoplasmic segment. Proc Natl Acad Sci USA. 1987; 84(24): 9145–9149. DOI: 10.1073/pnas.84.24.9145

[19] Koller BH, Geraghty DE, Shimizu Y, DeMars R, Orr HT. HLA-E. A novel HLA class I gene expressed in resting T lymphocytes. J Immunol. 1988; 141(3): 897–904.

[20] Rodgers JR, Cook RG. MHC class Ib molecules bridge innate and acquired immunity. Nat Rev Immunol. 2005; 5 (6): 459–471. DOI: 10.1038/nri1635.

[21] Lee N, Ishitani A, Geraghty DE. HLA-F is a surface marker on activated lymphocytes. Eur J Immunol. 2010; 40 (8): 2308–2318. DOI:10.1002/eji.201040348.

[22] Lee N, Llano M, Carretero M, Ishitani A, Navarro F, López-Botet M, Geraghty DE. HLA-E is a major ligand for the natural killer inhibitory receptor CD94/NKG2A. Proc Natl Acad Sci U S A. 1998;95(9):5199–5204. DOI: 10.1073/pnas.95.9.5199.

[23] Derré L, Corvaisier M, Charreau B, Moreau A, Godefroy E, Moreau-Aubry A, Jotereau F, Gervois N. Expression and release of HLA-E by melanoma cells and melanocytes: potential impact on the response of cytotoxic effector cells. J Immunol. 2006;177(5):3100–7. DOIi: 10.4049/jimmunol.177.5.3100.

[24] O'Callaghan CA. Molecular basis of human natural killer cell recognition of HLA-E (human leucocyte antigen-E) and its relevance to clearance of pathogen-infected and tumour cells. Clin Sci (Lond). 2000;99(1):9–17. DOI: 10.1042/cs0990009.

[25] Kochan G, Escors D, Breckpot K, Guerrero-Setas D. Role of non-classical MHC class I molecules in cancer immunosuppression. Oncoimmunology. 2013;2(11):e26491. DOI: 10.4161/onci.26491.

[26] Sullivan LC, Hoare HL, McCluskey J, Rossjohn J, Brooks AG. A structural perspective on MHC class Ib molecules in adaptive immunity. Trends Immunol. 2006;27(9): 413–420. DOI: 10.1016/j.it.2006.07.006.

[27] Llano M, Lee N, Navarro F, et al. HLA-E-bound peptides influence recognition by inhibitory and triggering CD94/NKG2 receptors: preferential response to an HLA-G-derived nonamer. Eur J Immunol. 1998; 28(9): 2854–2863. DOI: 10.1002/(SICI)1521-4141(199809)

[28] Foroni I, Couto AR, Bettencourt BF, Santos M, Lima M, Bruges-Armas J. HLA-E, HLA-F, and HLA-G — The Non-Classical Side of the MHC Cluster. Chapter 3, pages 61-109, HLA, and associated important diseases. Intech, Edited by Yongzhi Xi, https://www.intechopen.com/books/hlaand-associated-important-diseases.

[29] Carosella ED, Moreau P, Le Maoult J, Le Discorde M, Dausset J, Rouas-Freiss N. HLA-G molecules: from maternal-fetal tolerance to tissue acceptance. Adv Immunol. 2003;81:199–252. DOI: 10.1016/s0065-2776(03)81006-4.

[30] Le Rond S, Le Maoult J, Créput C, Menier C, Deschamps M, Le Friec G, Amiot L, Durrbach A, Dausset J, Carosella ED, Rouas-Freiss N Alloreactive CD4+, and CD8+ T cells express the immunotolerant HLA-G molecule in mixed lymphocyte reactions: in vivo implications in transplanted patients. Eur J Immunol. 2004;34(3):649–660. DOI.1002/eji.200324266.

[31] LeMaoult J, Rouas-Freiss N, Carosella ED. HLA-G5 expression by trophoblast cells: the facts. Mol Hum Reprod. 2005;11(10):719–722. DOI: 10.1093/molehr/gah224

[32] Morales PJ, Pace JL, Platt JS, Phillips TA, Morgan K, Fazleabas AT, Hunt JS. Placental cell expression of HLA-G2 isoforms is limited to the invasive trophoblast phenotype. J Immunol. 2003;171(11):6215–6224. DOI: 10.4049/jimmunol.171.11.6215

[33] Carosella ED, Favier B, Rouas-Freiss N, Moreau P, Lemaoult J. Beyond the increasing complexity of the immunomodulatory HLA-G molecule. Blood, 2008;111(10):4862–4870. DOI: 10.1182/blood-2007-12-127662.

[34] Clements CS, Kjer-Nielsen L, Kostenko L, Hoare HL, Dunstone MA, Moses E, Freed K, Brooks AG, Rossjohn J, McCluskey J. Crystal structure of HLA-G: a nonclassical MHC class I molecule expressed at the fetal-maternal interface. Proc Natl Acad Sci U S A. 2005; 102(9):3360–3365. DOI: 10.1073/pnas.0409676102.

[35] Ravindranath MH, Terasaki PI, Pham T, Jucaud V. The Monospecificity of Novel Anti-HLA-E Monoclonal Antibodies Enables Reliable Immunodiagnosis, Immunomodulation of HLA-E, and Upregulation of CD8+ T Lymphocytes. Monoclon Antib Immunodiagn Immunother. 2015; 34 (3): 135–153. DOI: 10.1089/mab.2014.0096.

[36] Ravindranath MH, Taniguchi M, Chen CW, Ozawa M, Kaneku H, El-Awar N, Cai J, Terasaki PI. HLA-E monoclonal antibodies recognize shared peptide sequences on classical HLA class Ia: relevance to human natural HLA antibodies. Mol Immunol. 2010; 47(5): 1121–1131. DOI: 10.1016/j.molimm.2009.10.024.

[37] Ravindranath MH, Pham T, El-Awar N, Kaneku H, Terasaki PI. Anti-HLA-E mAb 3D12 mimics MEM-E/02 in binding to HLA-B and HLA-C alleles: Web-tools validate the immunogenic epitopes of HLA-E recognized by the antibodies. Mol Immunol. 2011; 48(4): 423–430. DOI: 10.1016/j.molimm.2010.09.011.

[38] Ravindranath MH, Kaneku H, El-Awar N, Morales-Buenrostro LE, Terasaki PI. Antibodies to HLA-E in nonalloimmunized males: pattern of HLA-Ia reactivity of anti-HLA-E-positive sera. J. Immunol.2010;185(3): 1935–1948. DOI: 10.4049/jimmunol.1000424.

[39] Ravindranath MH, Selvan SR, Terasaki PI. Augmentation of anti-HLA-E antibodies with concomitant HLA-Ia reactivity in IFNγ-treated autologous melanoma cell vaccine recipients. J

Immunotoxicol. 2012;9(3):282–291. DOI: 10.3109/1547691X.2011. 645582.

[40] Ben Yahia H, Babay W, Bortolotti D, Boujelbene N, Laaribi AB, Zidi N, Kehila M, Chelbi H, Boudabous A, Mrad K, Mezlini A, Di Luca D, Ouzari HI, Rizzo R, Zidi I. Increased plasmatic soluble HLA-G levels in endometrial cancer. Mol Immunol. 2018;99:82–86. DOI: 10.1016/j.molimm.2018.04.007.

[41] Rizzo R, Gabrielli L, Bortolotti D, Gentili V, Piccirilli G, Chiereghin A, Pavia C, Bolzani S, Guerra B, Simonazzi G, Cervi F, Capretti MG, Fainardi E, Luca DD, Landini MP, Lazzarotto T.J Study of Soluble HLA-G in Congenital Human Cytomegalovirus Infection. Immunol Res. 2016;2016:3890306. DOI: 10.1155/2016/3890306.

[42] Alegre E, Rizzo R, Bortolotti D, Fernandez-Landázuri S, Fainardi E, González A. Some basic aspects of HLA-G biology. J Immunol Res. 2014;2014: 657625. DOI: 10.1155/ 2014/657625.

[43] Felício LP, Porto IO, Mendes-Junior CT, Veiga-Castelli LC, Santos KE, Vianello-Brondani RP, Sabbagh A, Moreau P, Donadi EA, Castelli EC. Worldwide HLA-E nucleotide and haplotype variability reveals a conserved gene for coding and 3′ untranslated regions. Tissue Antigens. 2014;83(2):82–93. DOI:10.1111/ tan.12283.

[44] Grimsley C, Ober C. Population genetic studies of HLA-E: evidence for selection. Hum Immunol. 1997;52(1): 33–40. doi: 10.1016/S0198-8859(96) 00241-8.

[45] Grimsley C, Kawasaki A, Gassner C, Sageshima N, Nose Y, Hatake K, Geraghty DE, Ishitani A. Definitive high-resolution typing of HLA-E allelic polymorphisms: Identifying potential errors in existing allele data. Tissue Antigens. 2002;60(3):206–212. doi: 10.1034/j.1399-0039.2002.600302.x.

[46] Geraghty DE. HLA-E allelic variants. Correlating differential expression, peptide affinities, crystal structures, and thermal stabilities. J Biol Chem. 2003;278(7):5082–5090. DOI: 10.1074/jbc.M208268200.

[47] Kraemer T, Blasczyk R, Bade-Doeding C. HLA-E: a novel player for histocompatibility. J Immunol Res. 2014; 2014: 352160. DOI: 10.1155/2014/ 352160.

[48] Iwaszko M, Bogunia-Kubik K. Clinical significance of the HLA-E and CD94/NKG2 interaction. Arch Immunol Ther Exp (Warsz). 2011; 59(5): 353–367. DOI: 10.1007/s00005 011-0137-y.

[49] Kraemer T, Celik AA, Huyton T, Kunze-Schumacher H, Blasczyk R, Bade-Doding C. 2015. HLA-E: Presentation of a Broader Peptide Repertoire Impacts the Cellular Immune Response—Implications on HSCT Outcome. Stem Cells International. 2015, Article ID 346714, 12 pages. dx.d oi.org/10.1155/2015/346714.

[50] Celik AA, Kraemer T, Huyton T, Blasczyk R, Bade-Döding C. The diversity of the HLA-E-restricted peptide repertoire explains the immunological impact of the Arg107Gly mismatch. Immunogenetics. 2016;68(1): 29–41. doi: 10.1007/s00251-015-0880-z.

[51] Goel R, Kabeerdoss J, Mohan H, Danda S, Jayaseelan V, Kumar TS, Jude J, Bacon P, Joseph G, Danda D.Soluble-HLA-E: A follow up biomarker in Takayasu arteritis, independent of HLA-E genotype. Int J Rheum Dis. 2017 Apr 19. DOI: 10.1111/1756-185X.13027.

[52] Wagner B, da Silva Nardi F, Schramm S, Kraemer T, Celik AA, Dürig J, Horn PA, Dührsen U, Nückel H, Rebmann V.HLA-E allelic genotype correlates with HLA-E plasma levels and predicts early progression in chronic lymphocytic leukemia. Cancer. 2017;123 (5):814–823. DOI: 10.1002/cncr.30427.

[53] Morandi F, Pozzi S, Carlini B, Amoroso L, Pistoia V, Corrias MV. Soluble HLA-G and HLA-E Levels in Bone Marrow Plasma Samples Are Related to Disease Stage in Neuroblastoma Patients. J Immunol Res. 2016;2016:7465741. DOI:10.1155/2016/7465741.

[54] Morandi F, Cangemi G, Barco S, Amoroso L, Giuliano M, Gigliotti AR, Pistoia V, Corrias MV. Plasma levels of soluble HLA-E and HLA-F at diagnosis may predict the overall survival of neuroblastoma patients. Biomed Res Int. 2013;2013:956878. DOI: 10.1155/2013/956878.

[55] Allard M, Oger R, Vignard V, Percier JM, Fregni G, Périer A, Caignard A, Charreau B, Bernardeau K, Khammari A, Dréno B, Gervois N. Serum soluble HLA-E in melanoma: a new potential immune-related marker in cancer. PLoS One. 2011;6(6):e21118. DOI: 10.1371/journal.pone.0021118.

[56] Coupel S, Moreau A, Hamidou M, Horejsi V, Soulillou JP, Charreau B. Expression and release of soluble HLA-E is an immunoregulatory feature of endothelial cell activation. Blood. 2007; 109(7): 2806–2814. DOI: 10.1182/blood-2006-06-030213.

[57] Marín R, Ruiz-Cabello F, Pedrinaci S, Méndez R, Jiménez P, Geraghty DE, Garrido F. Analysis of HLA-E expression in human tumors. Immunogenetics. 2003;54(11):767–775. DOI: 10.1007/s00251-002-0526-9.

[58] Gonçalves AS, Oliveira JP, Oliveira CF, Silva TA, Mendonça EF, Wastowski IJ, Batista AC.Relevance of HLA-G, HLA-E and IL-10 expression in lip carcinogenesis. Hum Immunol. 2016;77 (9):785-790. doi: 10.1016/j.humimm.2015.12.001.

[59] Silva TG, Crispim JC, Miranda FA, Hassumi MK, de Mello JM, Simões RT, Souto F, Soares EG, Donadi EA, Soares CP. Expression of the nonclassical HLA-G and HLA-E molecules in laryngeal lesions as biomarkers of tumor Invasiveness. Histol Histopathol. 2011; 26(12):1487-97. doi: 10.14670/HH-26.1487.

[60] van Esch EM, Tummers B, Baartmans V, Osse EM, Ter Haar N, Trietsch MD, Hellebrekers BW, Holleboom CA, Nagel HT, Tan LT, Fleuren GJ, van Poelgeest MI, van der Burg SH, Jordanova ES. Alterations in classical and non-classical HLA expression in recurrent and progressive HPV-induced usual vulvar intraepithelial neoplasia and implications for immunotherapy. Int J Cancer. 2014;135(4):830–842. DOI: 10.1002/ijc.28713.

[61] Djajadiningrat RS, Horenblas S, Heideman DA, Sanders J, de Jong J, Jordanova ES. Classic and nonclassic HLA class I expression in penile cancer and relation to HPV status and clinical outcome. J Urol. 2015;193(4):1245-51. doi: 10.1016/j.juro.2014.11.057.

[62] Mittelbronn M, Simon P, Löffler C, Capper D, Bunz B, Harter P, Schlaszus H, Schleich A, Tabatabai G, Goeppert B, Meyermann R, Weller M, Wischhusen J. Elevated HLA-E levels in human glioblastomas but not in grade I to III astrocytomas correlate with infiltrating CD8+ cells. J Neuroimmunol. 2007;189 (1-2):50-8.

[63] Kren L, Muckova K, Lzicarova E, Sova M, Vybihal V, Svoboda T, Fadrus P, Smrcka M, Slaby O, Lakomy R, Vanhara P, Krenova Z, Michalek J. Production of immune-modulatory nonclassical molecules HLA-G and HLA-E by tumor infiltrating ameboid microglia/macrophages in glioblastomas: a role in innate immunity? J Neuro immunol. 2010;220(1-2):131-5. doi: 10.1016/j.jneuroim. 2010.01.014.

[64] Kren L, Slaby O, Muckova K, Lzicarova E, Sova M, Vybihal V,

Svoboda T, Fadrus P, Lakomy R, Vanhara P, Krenova Z, Sterba J, Smrcka M, Michalek J. Expression of immune-modulatory molecules HLA-G and HLA-E by tumor cells in glioblastomas: an unexpected prognostic significance? Neuropathology. 2011;31(2):129-34. doi: 10.1111/j.1440-1789.2010.01149.x.

[65] Wolpert F, Roth P, Lamszus K, Tabatabai G, Weller M, Eisele G. HLA-E contributes to an immune-inhibitory phenotype of glioblastoma stem-like cells. J Neuroimmunol. 2012; 50(1-2):27-34. doi: 10.1016/j.jneuroim.2012.05.010.

[66] Wischhusen J, Friese MA, Mittelbronn M, Meyermann R, Weller M. HLA-E protects glioma cells from NKG2D-mediated immune responses in vitro: implications for immune escape in vivo. J Neuropathol Exp Neurol. 2005; 64(6):523-8.

[67] Bossard C, Bézieau S, Matysiak-Budnik T, Volteau C, Laboisse CL, Jotereau F, Mosnier JF. HLA-E/β2 microglobulin overexpression in colorectal cancer is associated with the recruitment of inhibitory immune cells and tumor progression. Int J Cancer. 2012;131(4) :855–863. DOI: 10.1002/ijc.26453.

[68] Costa Arantes DA, Gonçalves AS, Jham BC, Duarte ECB, de Paula ÉC, de Paula HM, Mendonça EF, Batista AC. Evaluation of HLA-G, HLA-E, and PD-L1 proteins in oral osteosarcomas. Oral Surg Oral Med Oral Pathol Oral Radiol. 2017 Jun;123(6):e188-e196. doi: 10.1016/j.oooo.2016.12.002.

[69] Mosconi C, Arantes DAC, Gonçalves AS, Alencar RCG, Oliveira JC, Silva TA, Mendonça EF, Batista AC. Immunohistochemical investigations on the expression of programmed cell death ligand 1, human leukocyte antigens G and E, and granzyme B in intraoral mucoepidermoid carcinoma. Arch Oral Biol. 2017 Nov;83:55-62. doi: 10.1016/j.archoralbio.2017.07.004.

[70] Reimer MS, Engels CC, Putter H, Morreau H, Liefers GJ, van de Velde CJ, Kuppen PJ. Prognostic value of HLA class I, HLA-E, HLA-G and Tregs in rectal cancer: a retrospective cohort study. BMC Cancer. 2014 Jul 5;14:486. doi: 10.1186/1471-2407-14-486.

[71] Levy EM, Bianchini M, Von Euw EM, Barrio MM, Bravo AI, Furman D, Domenichini E, Macagno C, Pinsky V, Zucchini C, Valvassori L, Mordoh J. Human leukocyte antigen-E protein is overexpressed in primary human colorectal cancer. Int J Oncol. 2008;32 (3):633–641.

[72] Levy EM, Sycz G, Arriaga JM, Barrio MM, von Euw EM, Morales SB, González M, Mordoh J, Bianchini M. Cetuximab-mediated cellular cytotoxicity is inhibited by HLA-E membrane expression in colon cancer cells. Innate Immun. 2009 Apr;15(2):91-100. doi: 10.1177/175342 5908101404.

[73] Benevolo M, Mottolese M, Tremante E, Rollo F, Diodoro MG, Ercolani C, Sperduti I, Lo Monaco E, Cosimelli M, Giacomini P. High expression of HLA-E in colorectal carcinoma is associated with a favorable prognosis. J Transl Med. 2011 Oct 27;9:184. doi: 10.1186/1479-5876-9-184.

[74] Zhen ZJ, Ling JY, Cai Y, Luo WB, He YJ. Impact of HLA-E gene polymorphism on HLA-E expression in tumor cells and prognosis in patients with stage III colorectal cancer. Med Oncol. 2013 Mar;30(1):482. doi: 10.1007/s12032-013-0482-2.

[75] Zeestraten EC, Reimers MS, Saadatmand S, Goossens-Beumer IJ, Dekker JW, Liefers GJ, van den Elsen PJ, van de Velde CJ, Kuppen PJ. Combined analysis of HLA class I, HLA-E and HLA-G predicts prognosis in colon cancer patients. Br J Cancer. 2014 Jan 21; 110(2):459-68. doi: 10.1038/bjc.2013.696.

[76] Guo ZY, Lv YG, Wang L, Shi SJ, Yang F, Zheng GX, Wen WH, Yang AG. Predictive value of HLA-G and HLA-E in the prognosis of colorectal cancer patients. Cell Immunol. 2015 Jan;293(1): 10-6. doi: 10.1016/j.cellimm.2014.10.003.

[77] Özgül Özdemir RB, Özdemir AT, Oltulu F, Kurt K, Yiğittürk G, Kırmaz C. A comparison of cancer stem cell markers and nonclassical major histocompatibility complex antigens in colorectal tumor and noncancerous tissues. Ann Diagn Pathol. 2016 Dec;25: 60-63. doi: 10.1016/j.anndiagpath.2016.09.012.

[78] Huang R, Zhang D, Li F, Xiao Z, Wu M, Shi D, Xiang P, Bao Z. Loss of Fas expression and high expression of HLA-E promoting the immune escape of early colorectal cancer cells. Oncol Lett. 2017 May;13(5):3379-3386. doi: 10.3892/ol.2017.5891.

[79] Stangl S, Gross C, Pockley AG, Asea AA, Multhoff G. Influence of Hsp70 and HLA-E on the killing of leukemic blasts by cytokine/Hsp70 peptide-activated human natural killer (NK) cells. Cell Stress Chaperones. 2008 Summer;13(2): 221-30. doi: 10.1007/s12192-007-0008-y.

[80] Chen A, Shen Y, Xia M, Xu L, Pan N, Yin Y, Miao F, Shen C, Xie W, Zhang J. Expression of the nonclassical HLA class I and MICA/B molecules in human hepatocellular carcinoma. Neoplasma. 2011;58(5):371-6.

[81] Talebian Yazdi M, van Riet S, van Schadewijk A, Fiocco M, van Hall T, Taube C, Hiemstra PS, van der Burg SH. The positive prognostic effect of stromal CD8+ tumor-infiltrating T cells is restrained by the expression of HLA-E in non-small cell lung carcinoma. Oncotarget. 2016 Jan 19;7(3):3477-88. doi: 10.18632/oncotarget.6506.

[82] de Kruijf EM, Sajet A, van Nes JG, Natanov R, Putter H, Smit VT, Liefers GJ, van den Elsen PJ, van de Velde CJ, Kuppen PJ. HLA-E and HLA-G expression in classical HLA class I-negative tumors are of prognostic value for the clinical outcome of early breast cancer patients. J Immunol. 2010;185 (12):7452–7459. DOI: 10.4049/jimmunol.1002629.

[83] da Silva GB, Silva TG, Duarte RA, Neto NL, Carrara HH, Donadi EA, Gonçalves MA, Soares EG, Soares CP. Expression of the Classical and Nonclassical HLA Molecules in Breast Cancer. Int J Breast Cancer. 2013;2013: 250435. doi: 10.1155/2013/250435.

[84] Gooden M, Lampen M, Jordanova ES, Leffers N, Trimbos JB, van der Burg SH, Nijman H, van Hall T. HLA-E expression by gynecological cancers restrains tumor-infiltrating CD8 T lymphocytes. Proc Natl Acad Sci U S A. 2011;108(26):10656-61. doi: 10.1073/pnas.1100354108.

[85] Gonçalves MA, Le Discorde M, Simões RT, Rabreau M, Soares EG, Donadi EA, Carosella ED. Classical and non-classical HLA molecules and p16 (INK4a) expression in precursors lesions and invasive cervical cancer. Eur J Obstet Gynecol Reprod Biol. 2008 Nov;141(1):70-4. doi: 10.1016/j.ejogrb.2008.06.010.

[86] Spaans VM, Peters AA, Fleuren GJ, Jordanova ES. HLA-E expression in cervical adenocarcinomas: association with improved long-term survival. J Transl Med. 2012 Sep 4;10:184. doi: 10.1186/1479-5876-10-184.

[87] Ferns DM, Heeren AM, Samuels S, Bleeker MCG, de Gruijl TD, Kenter GG, Jordanova ES. Classical and non-classical HLA class I aberrations in primary cervical squamous- and adenocarcinomas and paired lymph node metastases. J Immunother Cancer. 2016 Nov 15;4:78. doi: 10.1186/s40425-016-0184-3.

[88] Andersson E, Poschke I, Villabona L, Carlson JW, Lundqvist A, Kiessling R,

Seliger B, Masucci GV. Non-classical HLA-class I expression in serous ovarian carcinoma: Correlation with the HLA-genotype, tumor infiltrating immune cells and prognosis. Oncoimmunology. 2015 Jul 25;5(1):e1052213

[89] Zheng H, Lu R, Xie S, Wen X, Wang H, Gao X, Guo L. Human leukocyte antigen-E alleles and expression in patients with serous ovarian cancer. Cancer Sci. 2015 May;106(5):522-8. doi: 10.1111/cas.12641.

[90] Hanak L, Slaby O, Lauerova L, Kren L, Nenutil R, Michalek J. Expression pattern of HLA class I antigens in renal cell carcinoma and primary cell line cultures: methodological implications for immunotherapy. Med Sci Monit. 2009 Dec;15(12):CR638-43.

[91] Kren L, Valkovsky I, Dolezel J, Capak I, Pacik D, Poprach A, Lakomy R, Redova M, Fabian P, Krenova Z, Slaby O. HLA-G, and HLA-E specific mRNAs connote opposite prognostic significance in renal cell carcinoma. Diagn Pathol. 2012;29;7:58. DOI: 10.1186/1746-1596-7-58. PMID: 22640987.

[92] Zanetti BR, Carvalho-Galano DF, Feitosa NL, Hassumi-Fukasawa MK, Miranda-Camargo FA, Maciel LM, Ribeiro-Silva A, Soares EG. Differential expression of immune-modulatory molecule HLA-E in non-neoplastic and neoplastic lesions of the thyroid. Int J Immunopathol Pharmacol. 2013 Oct-Dec;26(4):889-96.

[93] Kren L, Fabian P, Slaby O, Janikova A, Soucek O, Sterba J, Krenova Z, Michalek J, Kral Z. Multifunctional immune-modulatory protein HLA-E identified in classical Hodgkin lymphoma: possible implications. Pathol Res Pract. 2012 Jan 15;208(1):45-9. doi: 10.1016/j.prp.2011.11.004.

[94] McWilliams EM, Mele JM, Cheney C, Timmerman EA, Fiazuddin F, Strattan EJ, Mo X, Byrd JC, Muthusamy N, Awan FT. Therapeutic CD94/NKG2A blockade improves natural killer cell dysfunction in chronic lymphocytic leukemia. Oncoimmunology. 2016 Sep 9;5(10): e1226720.

[95] Sensi M, Pietra G, Molla A, Nicolini G, Vegetti C, Bersani I, Millo E, Weiss E, Moretta L, Mingari MC, Anichini A. Peptides with dual binding specificity for HLA-A2 and HLA-E are encoded by alternatively spliced isoforms of the antioxidant enzyme peroxiredoxin 5. Int Immunol. 2009 Mar;21(3):257-68. doi: 10.1093/intimm

[96] Ravindranath, M.H., Jucaud, V., Ferrone, S. (2017) Monitoring native HLA-I trimer specific antibodies in Luminex multiplex single antigen bead assay: Evaluation of beadsets from different manufacturers. J Immunol Methods. 2017, 450:73–80. DOI: 10.1016/j.jim 201707. 016 PMID: 28782523. Corrigendum to Monitoring native HLA-I trimer specific antibodies in Luminex multiplex single antigen bead assay: Evaluation of beadsets from different manufacturers J. Immunol. Methods, 2018, 460:125. DOI:10.1016/j. jim. 2018. 07. 008.

[97] Ravindranath MH, Hopfield J, Ferrone S. HLA-E restricted monoclonal antibodies: Therapeutic potential as a double-edged sword against tumor progression. Internal Rev. Med. 2017;3 (12):1–49.

[98] Sasaki T, Ravindranath MH, Terasaki PI, Freitas MC, Kawakita S, Jucaud V. Gastric cancer progression may involve a shift in HLA-E profile from an intact heterodimer to β2m-free monomer. Int J Cancer. 2014;134(7): 1558–1570. DOI: 10.1002/ijc.28484.

[99] Ravindranath, M.H., Flippone, E. J., Mahowald, G., Callender, C, Babu, A., Saidman, S., Ferrone, S. (2018) Significance of the intraindividual

variability of HLA IgG antibodies in renal disease patients observed with different beadsets monitored with two different secondary antibodies on a Luminex platform. Immunologic Res. 66: 584–604. DOI: 10.1007/s12026-018-9027-2.

[100] Ravindranath, MH, Zhu D, Pham T, Jucaud V, Hopfield J, Kawakita S, Terasaki PI. Anti-HLA-E monoclonal antibodies reacting with HLA-Ia and Ib alleles like IVIg as potential IVIg-immunomimetics: an evolving therapeutic concept. Clin Transpl.2013: 293–305.

[101] Ravindranath, MH. HLA Class Ia and Ib Polyreactive Anti-HLA-E IgG2a Monoclonal Antibodies (TFL-006 and TFL-007) Suppress Anti-HLA IgG Production by CD19+ B Cells and Proliferation of CD4+ T Cells While Upregulating Tregs. J Immunol Res. 2017:3475926. DOI: 10.1155/ 2017/ 3475926.

[102] Hurez V, Kaveri SV, Mouhoub A, Dietrich G, Mani JC, Klatzmann D, Kazatchkine MD. Anti-CD4 activity of normal human immunoglobulin G for therapeutic use. (Intravenous immunoglobulin, IVIg). Ther Immunol. 1994;1(5):269–277.

[103] Amran D, Renz H, Lack G, Bradley K, Gelfand EW. Suppression of cytokine-dependent human T-cell proliferation by intravenous immunoglobulin. Clin Immunol Immunopathol. 1994 Nov;73(2):180–186. DOI: 10.1006/clin.1994.1186.

[104] Andersson J, Skansén-Saphir U, Sparrelid E, Andersson U. Intravenous immune globulin affects cytokine production in T lymphocytes and monocytes/macrophages. Clin Exp Immunol. 1996;104 Suppl 1:10–104 Suppl 1:20.

[105] Aktas O, Waiczies S, Grieger U, Wendling U, Zschenderlein R, Zipp F. Polyspecific immunoglobulins (IVIg) suppress proliferation of human (auto) antigen-specific T cells without inducing apoptosis. J Neuroimmunol. 2001 Mar 1;114(1–2):160–167. DOI: 10.1016/ s0165-5728(01)00243-0.

[106] MacMillan HF, Lee T, Issekutz AC. Intravenous immunoglobulin G-mediated inhibition of T-cell proliferation reflects an endogenous mechanism by which IgG modulates T-cell activation. Clin Immunol. 2009;132 (2):222–233. DOI: 10.1016/j. clim.2009.04.002.

[107] Aubin E, Lemieux R, Bazin R. Indirect inhibition of in vivo and in vitro T-cell responses by intravenous immunoglobulins due to impaired antigen presentation. Blood. 2010; 115 (9):1727–1734. DOI: 10.1182/blood-2009-06-225417.

[108] Kondo N, Ozawa T, Mushiake K, Motoyoshi F, Kameyama T, Kasahara K, Kaneko H, Yamashina M, Kato Y, Orii T. J Suppression of immunoglobulin production of lymphocytes by intravenous immunoglobulin. Clin Immunol. 1991;11(3):152–158. DOI: 10.1007/BF00918683.

[109] Kondo N, Kasahara K, Kameyama T, Suzuki Y, Shimozawa N, Tomatsu S, Nakashima Y, Hori T, Yamagishi A, Ogawa T, et al. Intravenous immunoglobulins suppress immunoglobulin productions by suppressing Ca(2+)-dependent signal transduction through Fc gamma receptors in B lymphocytes. Scand J Immunol. 1994 Jul;40(1):37–42. DOI: 10.1111/j.1365-3083.1994.tb03430.x.

[110] Glotz D, Haymann JP, Sansonetti N, Francois A, Menoyo-Calonge V, Bariety J, Druet P.Suppression of HLA-specific alloantibodies by high-dose intravenous immunoglobulins (IVIg). A potential tool for transplantation of immunized patients. Transplantation.

1993;56(2):335–337. DOI: 10.1097/00007890-199308000-00015.

[111] Glotz D, Antoine C, Haymann JP, Julia P, Duboust A, Bariéty J. Intravenous immunoglobulins and kidney transplantation in patients with anti-HLA antibodies. Adv Nephrol Necker Hosp. 2000;30:221–233.

[112] Tyan DB, Li VA, Czer L, Trento A, Jordan SC. Intravenous immunoglobulin suppression of HLA alloantibody in highly sensitized transplant candidates and transplantation with a histoincompatible organ. Transplantation. 1994;57(4):553–562.

[113] Jordan SC, Quartel AW, Czer LS, Admon D, Chen G, Fishbein MC, Schwieger J, Steiner RW, Davis C, Tyan DB.Posttransplant therapy using high-dose human immunoglobulin (intravenous gammaglobulin) to control acute humoral rejection in renal and cardiac allograft recipients and potential mechanism of action. Transplantation. 1998;66(6):800–805. DOI: 10.1097/00007890-199809270-00017.

[114] Okuda S, Kamei S, Sasaki T. Immunoglobulin G Enhances Generation of Inducible T Regulatory Cells and Increases Their Regulatory Function. Biol Pharm Bull. 2018;41(12): 1830–1836. DOI: 10.1248/bpb.b18-00548.

[115] Ephrem A, Chamat S, Miquel C, Fisson S, Mouthon L, Caligiuri G, Delignat S, Elluru S, Bayry J, Lacroix-Desmazes S, Cohen JL, Salomon BL, Kazatchkine MD, Kaveri SV, Misra N. Expansion of CD4+CD25+ regulatory T cells by intravenous immunoglobulin: a critical factor in controlling experimental autoimmune encephalomyelitis. Blood. 2008; 111(2): 715–722. DOI: 10.1182/blood-2007-03-079947.

[116] Zhu D, Ravindranath MH, Terasaki PI, Miyazaki T, Pham T, Jucaud V. Suppression of allo-human leucocyte antigen (HLA) antibodies secreted by B memory cells in vitro: intravenous immunoglobulin (IVIg) versus a monoclonal anti-HLA-E IgG that mimics HLA-I reactivities of IVIg. Clin Exp Immunol. 2014;177(2):464–477. DOI: 10.1111/cei.12307.

[117] Ravindranath MH, Terasaki PI, Pham T, Jucaud V, Kawakita S. Suppression of blastogenesis and proliferation of activated CD4(+) T cells: intravenous immunoglobulin (IVIg) versus novel anti-human leucocyte antigen (HLA)-E monoclonal antibodies mimicking HLA-I reactivity of IVIg. Clin Exp Immunol. 2014;178(1): 154–177. DOI: 10.1111/cei.12391.

[118] Ravindranath MH, Zhu D, Pham T, Jucaud V, Hopfield J, Kawakita S, Terasaki PI. Anti-HLA-E monoclonal antibodies reacting with HLA-la and lb alleles like IVIg as potential IVIg-immunomimetics: an evolving therapeutic concept. Clin Transpl. 2013: 293–305

[119] Hicklin DJ, Marincola FM, Ferrone S. HLA class I antigen downregulation in human cancers: T-cell immunotherapy revives an old story. Mol Med Today. 1999;5(4):178–186. DOI: 10.1016/s1357-4310(99)01451-3.

[120] Marincola FM, Jaffee EM, Hicklin DJ, Ferrone S. Escape of human solid tumors from T-cell recognition: molecular mechanisms and functional significance. Adv Immunol. 2000;74: 181–273. DOI: 10.1016/s0065-2776(08)60911-6.

[121] Campoli M, Chang CC, Oldford SA, Edgecombe AD, Drover S, Ferrone S. HLA antigen changes in malignant tumors of mammary epithelial origin: molecular mechanisms and clinical implications. Breast Dis. 2004;20:105–125. DOI: 10.3233/bd-2004-20112.

[122] Campoli M, Chang CC, Ferrone S. HLA class I antigen loss, tumor immune

escape, and immune selection. Vaccine. 2002;20 Suppl 4:A40–A45. DOI: 10.1016/s0264-410x(02)00386-9.

[123] Chang CC, Campoli M, Ferrone S. Classical and nonclassical HLA class I antigen and NK Cell-activating ligand changes in malignant cells: current challenges and future directions. Adv Cancer Res. 2005;93:189–234. DOI: 10.1016/S0065-230X(05)93006-6.

[124] Algarra I, García-Lora A, Cabrera T, Ruiz-Cabello F, Garrido F. The selection of tumor variants with altered expression of classical and nonclassical MHC class I molecules: implications for tumor immune escape. Cancer Immunol Immunother. 2004;53(10):904–10. DOI: 10.1007/s00262-004-0517-9.

[125] Cabrera T, López-Nevot MA, Gaforio JJ, Ruiz-Cabello F, Garrido F. Analysis of HLA expression in human tumor tissues. Cancer Immunol Immunother. 2003;52(1):1–9. DOI: 10.1007/s00262-002-0332-0.

[126] Seliger B. The non-classical antigens of HLA-G and HLA-E as diagnostic and prognostic biomarkers and as therapeutic targets in transplantation and tumors. Clin Transpl. 2013:465–472.

[127] Seliger B, Abken H, Ferrone S. HLA-G, MIC expression in tumors, and their role in anti-tumor immunity. Trends Immunol. 2003;24(2):82–87. DOI: 10.1016/s1471-4906(02) 00039-x.

[128] Bukur J, Jasinski S, Seliger B. The role of classical and non-classical HLA class I antigens in human tumors. Semin Cancer Biol. 2012;22(4):350–358. DOI: 10.1016/ j.semcancer.2012.03.003.

[129] Dutta N, Gupta A, Mazumder DN, Banerjee S. Down-regulation of locus-specific human lymphocyte antigen class I expression in Epstein-Barr virus-associated gastric cancer: implication for viral-induced immune evasion. Cancer. 2006;106(8):1685–1693. DOI: 10.1002/ cncr.21784.

[130] Dutta N, Majumder D, Gupta A, Mazumder DN, Banerjee S. Analysis of human lymphocyte antigen class I expression in gastric cancer by a reverse transcriptase-polymerase chain reaction. Hum Immunol. 2005;66(2): 164–169. DOI: 10.1016/j.humimm. 2004.10.010.

[131] Huang Z, Hyodo H, Fujii T, Nagamatsu T, Matsumoto J, Kawana K, Yamashita T, Yasugi T, Kozuma S, Taketani Y. Effect of progesterone on HLA-E gene expression in JEG-3 choriocarcinoma cell line. Am J Reprod Immunol. 2009;61(3):221–226. DOI: 10.1111/j.1600-0897.2008.00684.x.

[132] Bianchini M, Levy E, Zucchini C, Pinski V, Macagno C, De Sanctis P, Valvassori L, Carinci P, Mordoh J. Comparative study of gene expression by cDNA microarray in human colorectal cancer tissues and normal mucosa. Int J Oncol. 2006;29(1):83–94.

[133] Tremante E, Ginebri A, Lo Monaco E, Benassi B, Frascione P, Grammatico P, Cappellacci S, Catricalà C, Arcelli D, Natali PG, Di Filippo F, Mottolese M, Visca P, Benevolo M, Giacomini P. A melanoma immune response signature including Human Leukocyte Antigen-E. Pigment Cell Melanoma Res. 2014;27(1): 103–112. DOI: 10.1111/ pcmr.12164.

[134] Zhang X, Lin A, Zhang JG, Bao WG, Xu DP, Ruan YY, Yan WH. Alteration of HLA-F and HLA I antigen expression in the tumor is associated with survival in patients with esophageal squamous cell carcinoma. Int J Cancer. 2013;132(1):82–89. DOI: 10.1002/ ijc.27621.

[135] Petrie EJ, Clements CS, Lin J, Sullivan LC, Johnson D, Huyton T, Heroux A, Hoare HL, Beddoe T, Reid HH, Wilce MC, Brooks AG, Rossjohn J. CD94-NKG2A recognition of human

leukocyte antigen (HLA)-E bound to an HLA class I leader sequence. J Exp Med. 2008;205(3):725–735. DOI: 10.1084/jem.20072525.

[136] Braud VM, Allan DS, O'Callaghan CA, Söderström K, D'Andrea A, Ogg GS, Lazetic S, Young NT, Bell JI, Phillips JH, Lanier LL, McMichael AJ. HLA-E binds to natural killer cell receptors CD94/NKG2A, B and C. Nature. 1998; 391(6669):795–799. DOI: 10.1038/35869.

Chapter 4

Analytical Characterization of Monoclonal Antibodies with Novel Fc Receptor-Based Chromatography Technique

Atis Chakrabarti, Jukka Kervinen, Egbert Müller, Toru Tanaka and Kazuaki Muranaka

Abstract

Most clinically approved large biotherapeutics are monoclonal antibodies (mAbs), primarily belonging to immunoglobulin G subclass-1 (IgG1) and, to a lesser extent, IgG2 and IgG4. Glycosylation is the main source of post-translational heterogeneity of mAbs, impacting their drug therapeutic mechanism of action (MOA). Glycosylation is also one of the critical factors in drug product solubility, kinetics, stability and efficacy. Thus, monitoring glycan critical quality attributes (CQAs) is an essential part of any biopharmaceutical development. The binding affinity of an IgG to its cellular Fc receptor (FcR) depends on both its IgG subclass and Fc domain glycosylation pattern. Since composition of the N-glycans also correlates to the Antibody-Dependent Cellular Cytotoxicity (ADCC), the glycosylation pattern needs to be monitored for consistency in potency and efficacy. This applies for the original mAb biologics as well as biosimilars. In this chapter, we present a truly novel way to assess the variances in mAb glycoforms using FcγRIIIa-based affinity chromatography. First, a brief overview of the Fc receptor function is presented. Then, the principle of FcR-based affinity chromatography is explained including how this column's potential to analyze a variety of mAbs according to their N-glycan content is highly selective and robust. Finally, we provide examples of the FcR column's potential to improve analytical characterization of mAbs with practical applications such as effective cell line screening, monitoring of glycoengineering, process development and process control in manufacturing.

Keywords: FcR, glycoform, N-glycan, monoclonal antibody, mAb, biosimilar, affinity chromatography, antibody-dependent cellular cytotoxicity, ADCC

1. Introduction

Affinity chromatography is a popular method for the purification of biomolecules. The purification involves interaction of the biomolecule with a ligand covalently immobilized to a solid stationary phase. Elution occurs as a function of binding strength of the biomolecule to the stationary phase, tighter the binding later the elution time in a linear gradient. Due to high selectivity and fast separation, affinity chromatography is regarded as the most widely used purification method for capture step in biopharmaceutical industry. Different types of affinity chromatography ligands are available as resins and pre-packed columns for purification at various scales. Similarly, different types of U/HPLC affinity chromatography columns are available for analytical characterization and quality control. This chapter focuses on a recently introduced novel FcR-based affinity column, TSKgel-FcR-IIIA-NPR, that enables chromatographic characterization of mAbs based on their N-glycan content attached to a highly conserved Asn-297 in Fc region.

2. Brief overview of Fc receptor

A brief overview of the Fc Receptor (FcR) structure and function is provided to best understand the chromatography principle of the column discussed in this chapter. FcR proteins belong to immunoglobulin (Ig) superfamily [1]. Interest in FcRs for biotherapeutic research has gained momentum since 1980s. The purification of FcR from the glycoprotein fraction of the placental membranes by chromatography was reported in 1982 [2]. A functional 40 kDa FcR, with low affinity for native IgG, was purified from the human peripheral nerve extract using $F(ab)_2$ fragments of mAb against placental FcR as affinity agent in 1989 [3]. Important role of FcR in IgG distribution to the brain [4], inhibition of cell activation [5] and in enhancement and suppression of the effector function [6] have also been reported. Overall, it became evident that FcRs are important for numerous biological functions.

2.1 FcγRs

The FcRs binding to immunoglobulin G (IgG) are known as Fc-gamma receptors (FcγR). FcγRs play essential role in immunity, inflammatory and infectious diseases [7]. Immune enhancement and suppression are influenced by binding to these FcγRs [6]. Additional interaction between hyaluronic acid (HA) and sialic acids on immune cells helps to optimize the FcR-mediated effector function [8]. Fcγ receptors do not bind to IgA or IgM [9].

2.2 Fcγ receptor binding to IgG

Typical IgG is Y-shaped protein of ~150 kDa in size, containing two heavy chains and two light chains (**Figure 1**). The heavy chain (HC) contains three constant domains (CH1–CH3) and a variable domain (VH) with three complementarity-determining regions (CDRs). The light chain (LC) has only one constant domain (CL) and a variable domain (VL) with CDRs. The Fd consists of VH and CH1. LC and Fd together form the antigen binding fragment (Fab). The CH2 domain of each heavy chain of IgG has a highly conservative asparagine (N) residue at position 297 (Asn297 or N297) that is almost invariably glycosylated. Fcγ receptor binding site is located near the hinge region of IgG, close to N297 in the CH2 domain. The most flexible portion of the hinge region is between CH1 and CH2 domains of a heavy chain. The four chains are covalently connected via disulfide bridges [10]. Fraction crystallizable (Fc) is composed of CH2 and CH3 domains of the two heavy chains. The highly conserved glycan moiety at position N297 infers structural changes to the Fc-region required for binding to FcγR. Subtle differences in the glycan composition at this site, thus, can affect the conformational rigidity of the Fc-structure, and may also alter the interaction with FcγR by direct contact [11].

2.3 FcγR subclasses

FcγRs are divided into three subclasses, abbreviated as FcγRI, FcγRII and FcγRIII. Extracellular regions of all the FcγRs are extremely homologous, whereas the

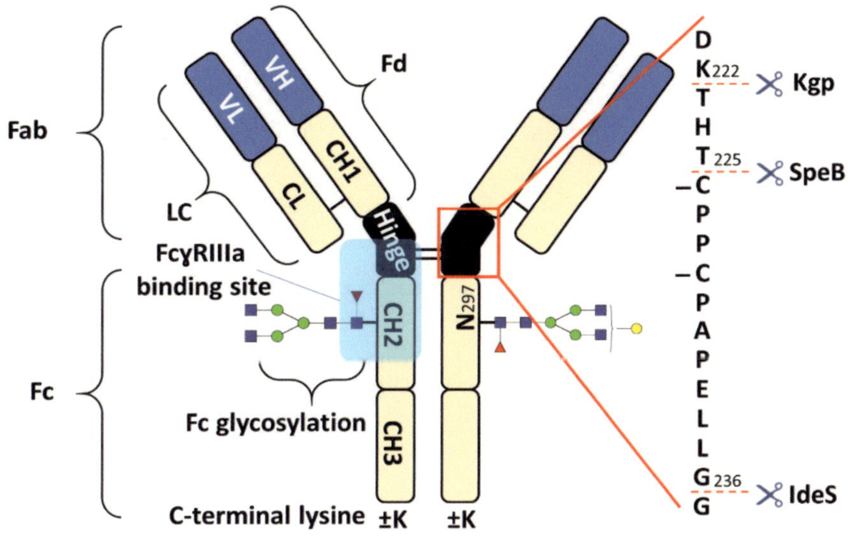

Figure 1.
Schematic overview of human IgG1 with a detailed sequence view into the hinge region [10] (reprinted with permission).

cytoplasmic domains differ considerably from each other [12]. FcγRI exhibits the highest affinity for IgG, K_a 10^8–10^9 M^{-1} whereas FcγRII and FcγRIII show a weaker affinity [13] for monomeric IgG, $K_a \leq 10^7$ M^{-1}. Receptor clustering is essential for FcγR signaling. FcγRIII (also known as CD16) is a cluster of differentiation molecule found on the surface of natural killer (NK) cells, neutrophils, monocytes and macrophages.

2.4 FcγRIII isoforms

FcγRIII exists in two different isoforms, (a) FcγRIIIa or CD16a and (b) FcγRIIIb or CD16b. Both forms take part in intracellular signal transduction. Two nearly identical genes in human encode these two isoforms. FcγRIIIa is a 50–65 kDa type-1 transmembrane protein whereas FcγRIIIb is a 48 kDa glycosylphosphatidylinositol (GPI)-anchored protein. This chapter focuses on the modified recombinant FcγRIIIa protein ligand, immobilized on a polymethacrylate stationary phase and packed into an analytical chromatography column that can be used for characterization of antibodies based on their N-glycan content on Asn297.

2.5 FcγRIIIa and glycosylation mode of IgG

Post-translational modifications, particularly glycosylation, of both IgG antibodies and Fcγ receptors modulate the affinity of their interaction. N-glycan (**Figure 2**) is a well-defined complex biantennary structure composed of a core hepta-saccharide, made up of N-acetylglucosamine (GlcNAc) and mannose, followed by variable additions of galactose, sialic acid (N-acetylneuraminic acid), fucose, and bisecting GlcNAc residues [11].

The attached glycans play various crucial roles on the function of immunoglobulins. Fc sialylation prolongs serum half-life of therapeutic antibodies [14]. **Figure 3** shows details of the structure of the glycosylated Fc fragment complexed to a FcγRIIIa receptor [15]. In non-fucosylated mAb, the carbohydrate-carbohydrate interactions increase binding affinity between N-Glycan of IgG-Fc and N-Glycan of FcR ($K_D = 7.2 \times 10^{-9}$ M) while in fucosylated mAb the carbohydrate-carbohydrate interaction is weekened or non-existent depending on the extent of steric hindrance of the fucose moiety ($K_D = 3.0 \times 10^{-7}$ M) [15].

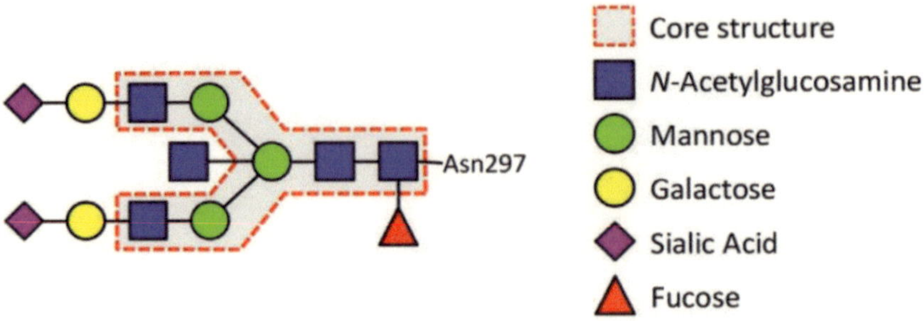

Figure 2.
A schematic representation of a common N-glycan structure where blue squares denote to GlcNAc, green circles denote to mannose, yellow circles denote to galactose, purple diamonds denote to sialic acid (N-acetylneuraminic acid) and red triangle denote to fucose (reprinted with permission).

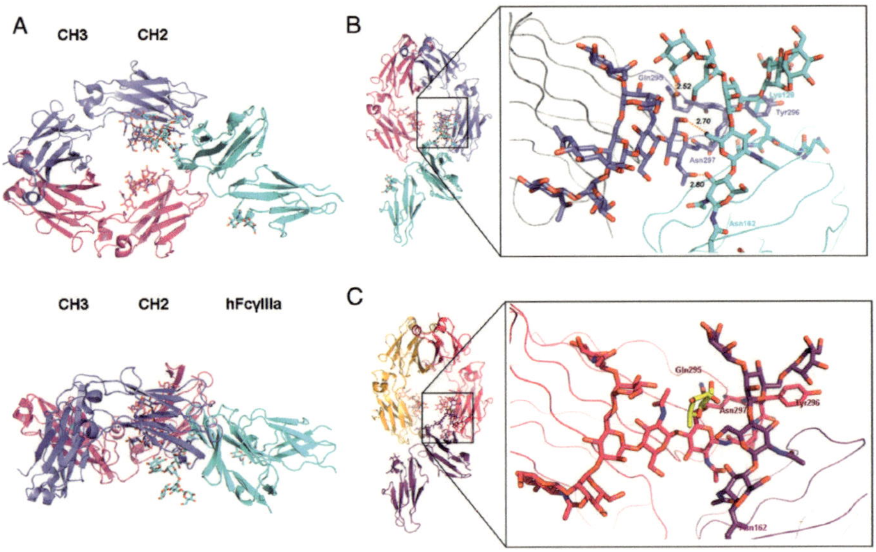

Figure 3.
Crystal structure of glycosylated Fc-FcγRIIIa complex. (A) Top and side views of the structure of the glycosylated Fc-FcγRIIIa complex. The Fc chains are shown in blue and magenta and the receptor in cyan. The oligosaccharides are depicted as ball-and-stick representation. (B) View on the interaction interface between afucosylated Fc fragment and glycosylated Fc receptor. Chain A of the Fc fragment is shown in blue, the Fc receptor in cyan. Hydrogen bonds are presented as dashed lines with distance between donor and acceptor atoms. (C) View on the interaction interface between fucosylated Fc fragment and glycosylated Fc receptor. Chain A of the Fc fragment is shown in magenta, the Fc receptor in dark violet. Core fucose of fucosylated Fc is highlighted in yellow [15]. (Reprinted with permission).

The glycosylation of Fc part is prerequisite for its affinity to FcR. Therapeutic monoclonal antibodies recognize specific cell surface-expressed antigens in malfunctioning cells (e.g. cancer cells) and elicit immune effector functions such as Antibody-Dependent Cell-mediated Cytotoxicity (ADCC) (**Figure 4**). Glycan composition at conserved N297 in IgG largely affects the binding affinity, thus regulating ADCC activity.

FcγRIIIa protein contacts Fc portion of IgG (to both CH2 regions) and to attached glycans. The glycans from FcγRIIIa side are also apparently in contact with glycans from IgG-Fc. The binding is asymmetrical in nature as revealed from the co-crystal structure of Fc-FcγRIII complex [16], although the stoichiometry of binding is 1:1. Lack of fucosylation in core Fc-glycan dramatically increases the ADCC activity due to enhanced binding affinity of FcγRIIIa to IgG [17].

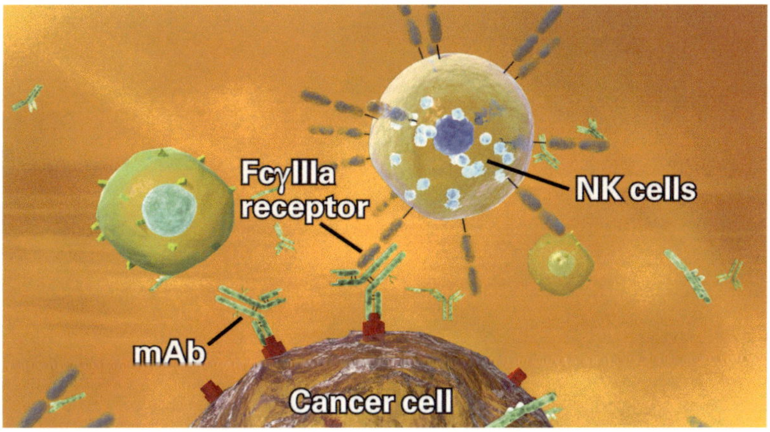

Figure 4.
Schematic representation of the antibody-Fc receptor interaction resulting in ADCC activity.

3. TSKgel FcR-IIIA-NPR affinity column

TSKgel FcR-IIIA-NPR affinity column contains non-porous polymethacrylate base beads as stationary phase. The ligand is a modified recombinant non-glycosylated FcγRIIIa of 20 kDa, produced in *E. coli* expression system. Recombinant FcγRIIIa ligand has eight amino acid substitutions as compared to its wild-type form. These changes were necessary for stabilization of the ligand structure [18]. The 2.7 Å crystal structure of recombinant FcγRIIIa protein verifies the molecular basis of the IgG-FcR complex formation. No significant difference was found between the crystal structures of glycosylated wild-type FcγRIIIa expressed in human embryonic kidney (HEK) cells vs. non-glycosylated recombinant FcγRIIIa from *E. coli* (**Figure 5**). This confirms suitability of *E. coli*-produced non-glycosylated FcγRIIIa to be used as an affinity ligand in a chromatography resin. Notably, no direct contact of the terminal Asn297 N-glycan galactose of IgG and modified non-glycosylated FcγRIIIa was observed [19]. Both proteins were crystallized as complex with Fc [19].

The dimension of the TSKgel FcR-IIIA-NPR column is 4.6 mm ID × 7.5 cm (l) with a total bed volume of 1.25 mL. Polymethacrylate-based matrix is composed of non-porous material with ~5 μm particle size. Maximum pressure limit of the column is 90 bar (9 MPa). The column is suitable for both HPLC and UHPLC instrument settings. The recommended run temperature is 15–25°C. The operational pH range is from pH 4 to 8. In general, most monoclonal antibodies bind effectively on the column at pH 6.5. Typically, 50 mM ammonium citrate (or ammonium acetate) buffer is used. A linear pH gradient from pH 6.5 to 4.5 over 16 column volumes (CV) at the flow rate of 1.0 mL/min is recommended. **Figure 6** shows a typical three-peak chromatographic profile for a monoclonal antibody in these settings. Sodium chloride can be added to buffer to enhance the separation if needed. Longer retention time indicates stronger mAb affinity to the ligand. However, as to the general composition of glycans, it should be noted that all the three peaks still contain a mixture of glycoforms with variable amounts of galactose and other sugar molecules (**Figure 7**).

From a related experiment (**Figure 7**), the three peaks were collected for glycan analysis and ADCC activity assay. Determination of the glycan structures revealed that the retention time increase correlates with increased number of the terminal galactose. Terminal galactose tends to stabilize conformation of the Fc region, providing tighter binding onto FcγRIIIa affinity ligand [20]. However, FcR column is not designed for quantitation of only galactose but to obtain a more general understanding of the variation in distribution of the glycan content among the

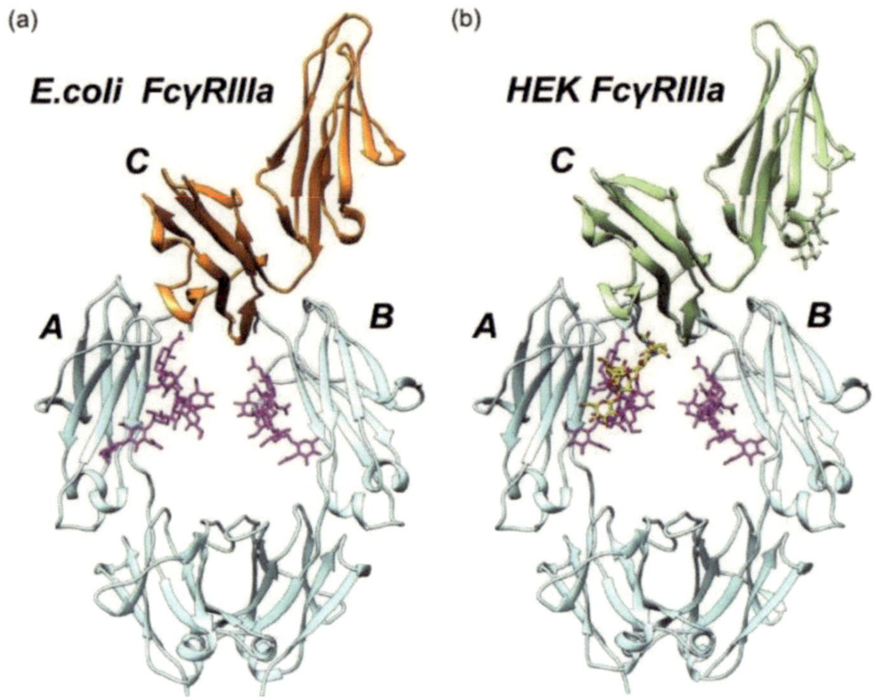

Figure 5.
Crystal structures of a recombinant non-glycosylated FcR ligand (Panel a) and a glycosylated native form (Panel b) (reprinted with permission).

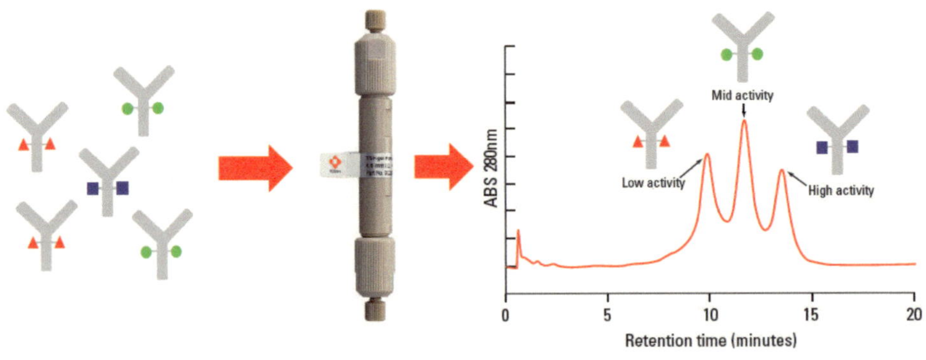

Figure 6.
A typical 3-peak elution profile for a monoclonal antibody using a TSKgel FcR-IIIA-NPR column with correlation to ADCC activity.

three peaks. Although added galactose increases retention time, other factors (as explained below) also affect the mAb binding affinity.

The results shown in the **Figure 7** support the binding model presented for galactose in the **Figure 8**. The crystal structure, basis of a cartoon model in the **Figure 8**, surprisingly did not show direct contact of the galactose units with the receptor that could more easily explain galactose effect on the affinity for FcγRIIIa. Instead, based on the evidence reported in literature so far, it has been proposed that the galactose moiety can influence the dynamic and conformational assembly of IgG-Fc. Hydrogen-deuterium exchange mass spectrometry

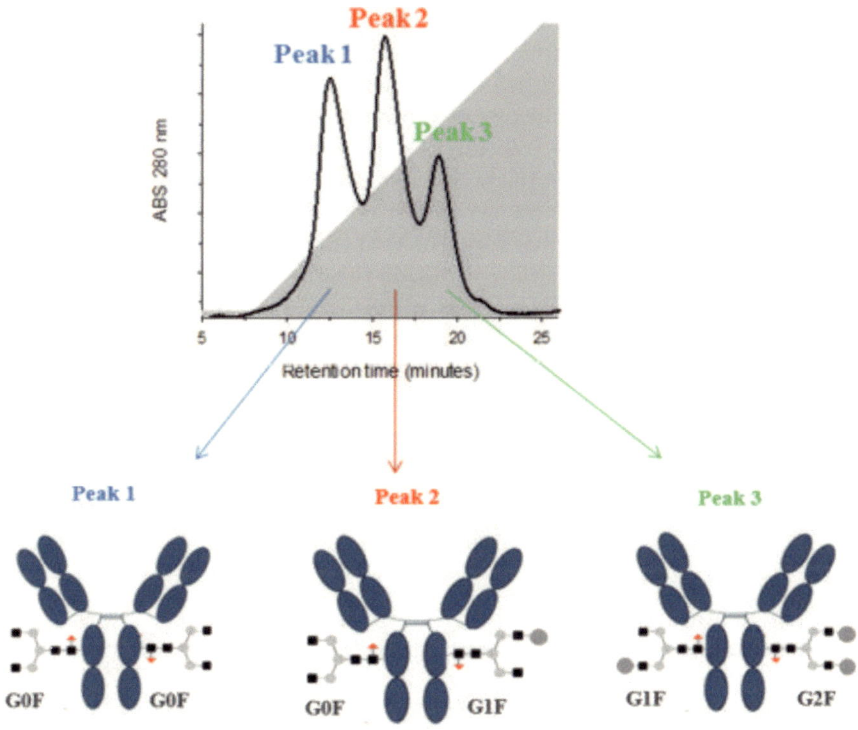

Figure 7.
Correlation between the number of galactose units and retention time [20] (reprinted with permission).

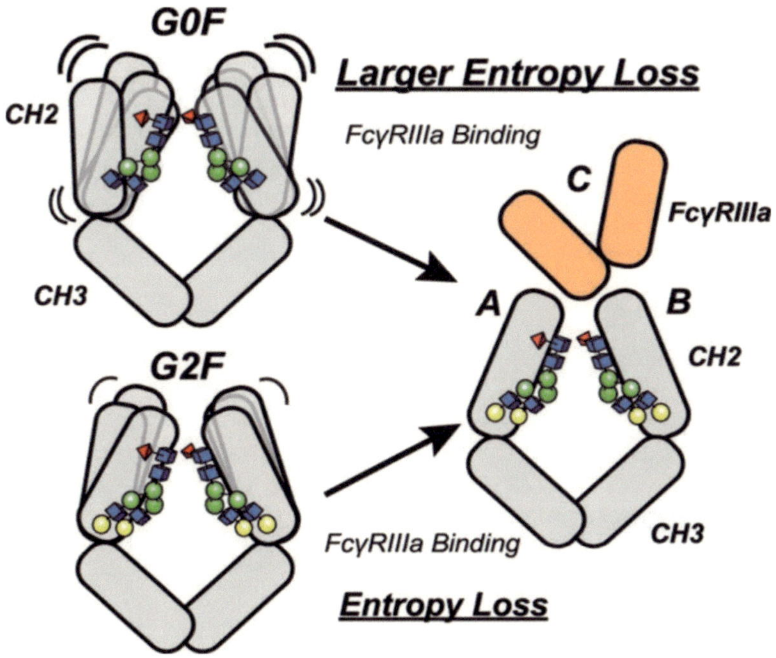

Figure 8.
Conformational entropy modulated by galactose content controls the binding affinity of IgG to FcγRIIIa [19]. (reprinted with permission).

(HDX-MS) study using purified IgG glycovariants support this hypothesis. By the deuterium exchange mass spectrometry, it was noticed that the deuterium uptake increases in the peptide ranging from 245 P to 256 T in the following order: Peak 1 > Peak 2 > Peak 3. This result implicates that this particular peptide exhibits a more rigid conformation as the fraction of galactose units increase. Differential Scanning Calorimetry (DSC) experiment also proved that the peak 3 contains antibodies with the highest galactose content, and it exhibited the greatest denaturation enthalpy. This result thus suggests that the terminal galactose engages in non-covalent interaction with surrounding residues leading to increased conformational stability. The value of entropy change decreased as the content of galactose increased, suggesting a reduction of the conformational entropy of the antibody. More specifically, terminal galactose moiety seems to especially stabilize the mAb hinge region. In N-glycans containing galactose, the CH2 domain remains in more rigid conformation as compared to the agalactosylated (G0F) glycoform (i.e. no galactose). Overall, the number of terminal galactoses have the greatest impact on the binding affinity of mAb onto the column [19]. However, the other types of glycans such as fucose, mannose and sialic acid also affect the binding.

4. Correlation between mAb retention time and ADCC activity

In order to confirm correlation with ADCC activity and retention time, Rituximab as an example [20], was analyzed and three successive peaks from TSKgel FcR-IIIA-NPR column were collected. ADCC bioassay was performed for these peaks using the Promega ADCC reporter assay kit (**Figure 9**). In the assay, higher luminescence (RLU units), as compared to mAb concentration, denotes to stronger ADCC activity. As a measure of ADCC potency, EC50 values (mAb concentration with 50% of the maximal ADCC activity) were determined. RLU units for Rituximab sample prior to load and for three fractions were plotted as a function of their concentrations (μg/mL) (**Figure 9**, Panel B). The order of ADCC activity in the peaks is as follows: peak 3 > peak 2 > mAb > peak 1. The result thus clearly indicates that the peak 3 has the highest ADCC activity as compared to the other two peaks. This study proves that the increased retention

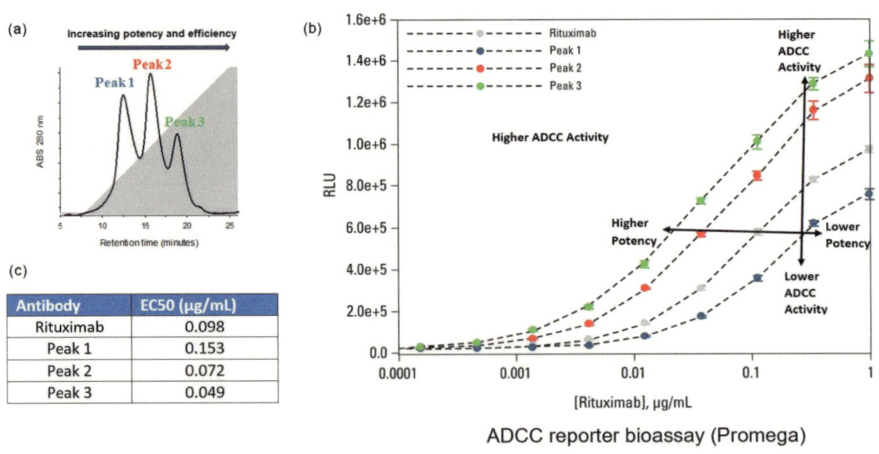

Antibody	EC50 (μg/mL)
Rituximab	0.098
Peak 1	0.153
Peak 2	0.072
Peak 3	0.049

Figure 9.
ADCC reporter bioassays for Rituximab and its glycoforms separated on TSKgel FcR-IIIA-NPR. Panel A: A typical 3-peak elution pattern for the mAb. Panel B: ADCC reporter bioassay for the mAb control and collected peaks from panel A. Panel C: EC50 values for the three peaks. (reprinted with permission).

time for the peaks obtained from the U/HPLC assay indeed correlates with the higher ADCC activity.

The core fucose can also exert a strong modulatory effect on the affinity for FcγRIIIa. **Figure 10** illustrates the location of a fucose molecule in the glycan moiety. In the mAb, the core fucose inhibits the carbohydrate contacts and decreases the binding affinity. Fucosylation of N-glycan thus reduces the affinity of mAb-FcR interaction by steric hindrance within the Fc cavity, obstructing the carbohydrate–carbohydrate interaction [15].

Afucosylated mAb binds effectively to FcγRIIIa ligand [19]. Subsequently, fucosylated and non-fucosylated mAbs were compared on the column where it was shown that deletion of core fucose significantly increased both retention time and the ADCC activity (**Figure 11**).

Sialic acid (N-acetylneuraminic acid) has a role in ADCC activity. Sialylation in the context of core fucosylation significantly decreased ADCC activity [21] as the sialic acid lowers the binding affinity to Fc receptor [19]. In absense of core fucosylation, sialylation doesn't have any significant impact on ADCC activity [21]. FcγRIIIa affinity chromatography yields longer retention by core fucosylation, terminal galactosylation and sialylation [18]. Overall, the effect

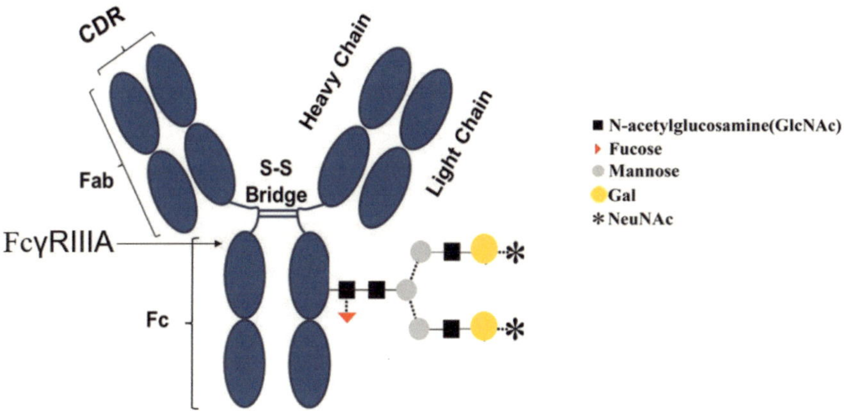

Figure 10.
Schematic representation of the core fucose location in the N-glycan moiety (reprinted with permission).

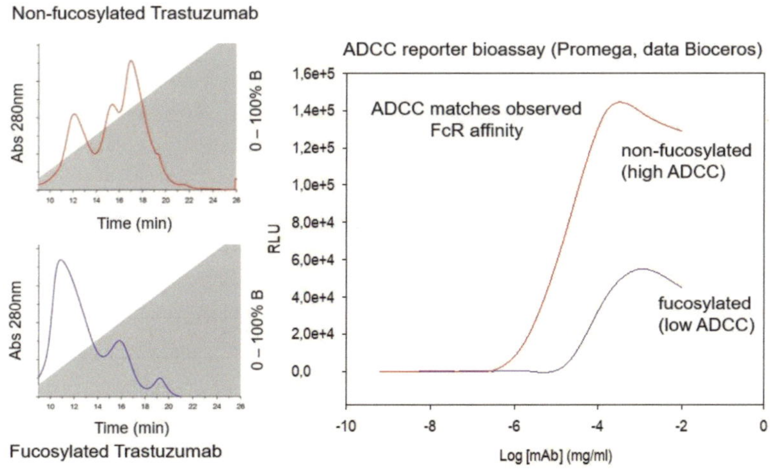

Figure 11.
Effect of the core fucose on ADCC activity [20] (reprinted with permission).

of different glycan molecules on the binding strength to Fc receptor can be arranged in the following order: Galactosylated (terminal) > Afucosylated > Sialylated (terminal) N-glycans in mAb. Similalry, retention time and ADCC activity is expected to be in the following order in the other forms of glycosylation patterns; A2G2 > A2G2S2 > A2G0, High mannose (HM) > FA2G2 and FA2G2S2 unless otherwise affected by any other factor. High mannose and A2G0 may be of similar activity. Complement-dependent cytotoxicity (CDC) is not significantly related to sugar structure [18].

5. Glycosylation profiling of a variety of monoclonal antibodies

IgG1 is the most abundant antibody in human body. Other antibodies are IgG2, IgG3, and IgG4 (**Figure 12**). Generally, IgG1, IgG3, and to some extent, IgG4 are formed against protein antigens. IgG2 is the major subclass formed against repetitive T cell-independent polysaccharide structures found on encapsulated bacteria [22]. Most of the biotherapeutics predominantly belong to IgG1 subclass. The antibodies from subclasses other than IgG1, as well as numerous engineered forms, are also gaining plenty of interest for use as biotherapeutics (**Figure 12**).

Several commercially available IgGs were recently analyzed in-house using TSKgel FcR-IIIA-NPR column to compare their elution profiles. Almost all IgGs yielded the typical 3-peak separation profile although there was substantial difference in each peak height between IgGs. Generally, IgG1, IgG3 and IgG4 subclasses bind to TSKgel FcR-IIIA-NPR whereas IgG2 does not have affinity. Due to

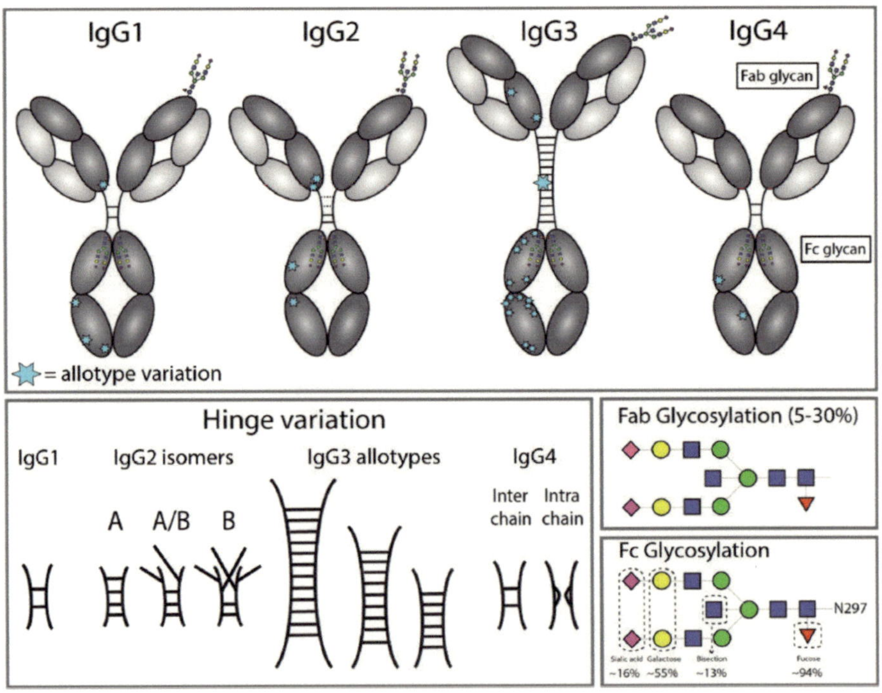

Figure 12.
A structural representation of the IgG subclasses and the variation within these subclasses, including allotypes, hinge variation, and glycosylation. The variation originating from allotypic polymorphisms in the immunoglobulin heavy gamma (IGHG) Fc domain is indicated with blue stars. Except for the star representing the variation in hinge length between IgG3 allotypes, each smaller blue star indicates amino acid variation at one particular residue in the constant domain. Glycans attached to N297 in Fc region are highly variable and the frequency of each glycan moiety on IgG antibodies in human serum is indicated [22] (reprinted with permission).

individual peak size differences, each mAb is indicative of its unique separation profile (**Figure 13**). This study thus shows that TSKgel FcR-IIIA-NPR column can be used for the analysis of a variety of IgG subclasses for glycosylation profiling. The results typically well correlate to ADCC activity.

Since the first FDA approvals of biosimilars in the USA in 2015, the interest continues to increase toward biosimilars. They are, in general, less costly to develop than the original or innovators. However, the biosimilar manufacturers are required to confirm the extent of similarity with the corresponding innovator. In the recent literature report [23], it was indeed confirmed that glycan microheterogeneity may play a critical role in effector function between the originals and biosimilars. In the study, it was shown that a biosimilar had a higher level of afucosylated glycans, resulting in a stronger FcγRIIIa binding affinity and increased ADCC activity. The study in the **Figure 14** also shows that the TSKgel FcR-IIIA-NPR column yielded dissimilar chromatographic pattern for biosimilars as compared to its innovator. Thus this column can be used for monitoring biosimilar consistent with the innovator biomolecule.

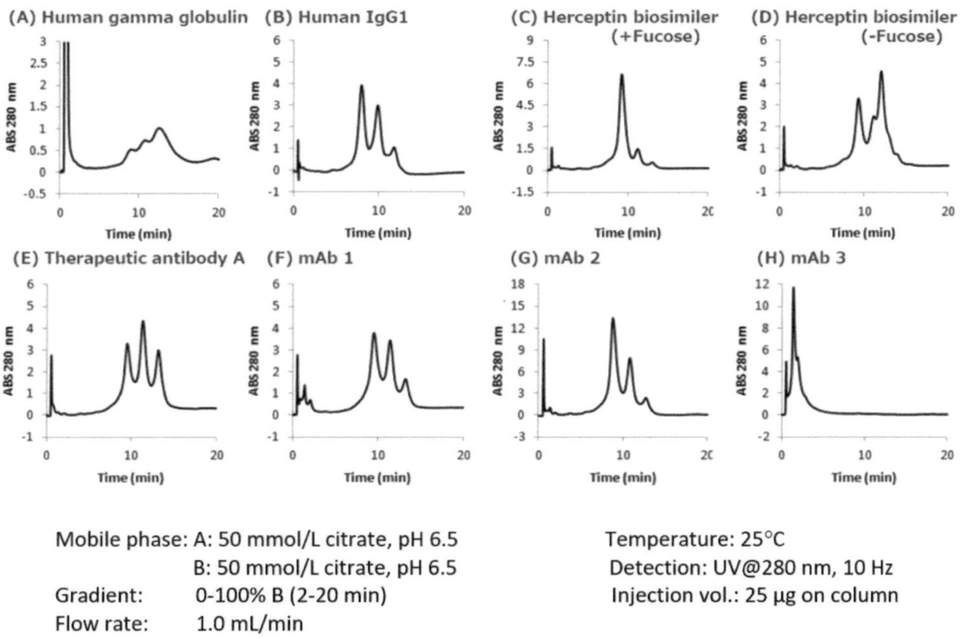

Figure 13.
Analysis of a variety of monoclonal antibodies on TSKgel FcR-IIIA-NPR.

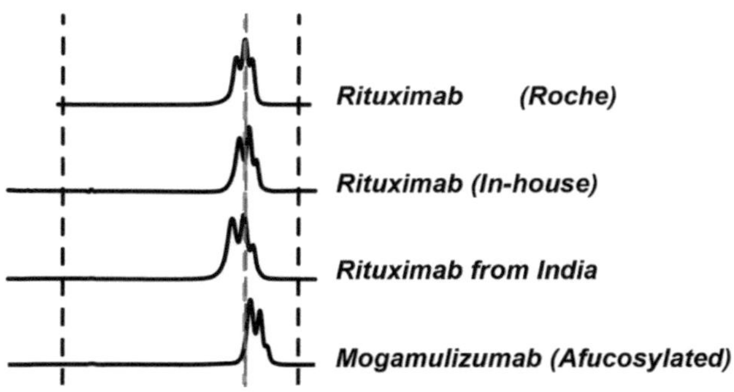

Figure 14.
Comparison of Roche's innovator Rituximab to its two biosimilar biotherapeutic forms. The figure also includes mogamulizumab (Poteligeo™) as an example of a completely afucosylated mAb.

6. Factors affecting the chromatographic separation

6.1 Salt concentration

Salts affect the separation of mAb on TSKgel FcR-IIIA-NPR. To best control the pH in the linear gradient, mobile phase should consist of a buffer with suitable buffering capacity whereas neutral salts are used to increase the ionic strength. Both components affect the binding affinity. Buffer provides pH control and salt ions provide charge shielding or stoichiometric ion bonding on the stationary phase and mAbs. Salts impart specific or non-specific effects by modulating protein–protein and protein-surface interactions. Binding affinity to the column depends on the binding constant. Increasing salt concentration have shown to lead to the elution at earlier retention time (**Figure 15**, Panel A), although the intensity of the effect probably is also related to the individual mAb studied. The binding strength is also dependent on the buffer used such as sodium citrate or sodium acetate. Citrate yielded stronger binding and hence higher retention time. Acetate buffer instead yielded better resolution of the peaks as compared to citrate (**Figure 15**, Panel B).

In the affinity chromatography, the optimum flow rate of elution may be dependent on the molecule-specific interaction with the ligand. Irrespective of the flow rates (**Figure 16**), all the three glycoform peaks eluted within 67% of mobile phase B when the analysis was carried out using a linear gradient of 50 mM sodium citrate buffer from pH 6.5 to 4.5 over 50 minutes at 20°C. Although flow rate did not have effect on elution pH, lower flow rate may be used to increase the sensitivity due to longer residence time in the flow cell.

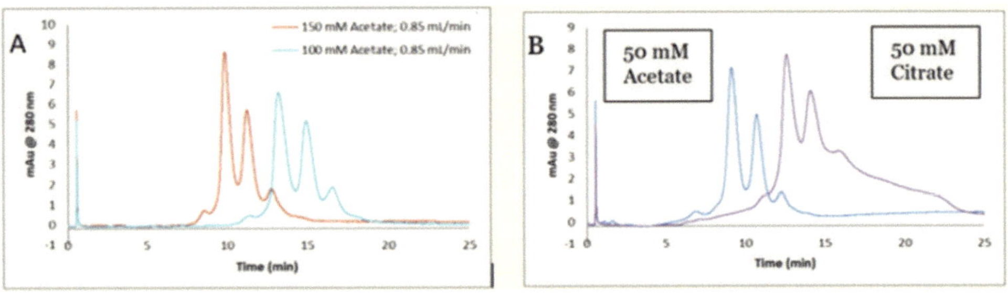

Figure 15.
Effect of salt (Panel A) and buffer (Panel B) on the separation of mAb glycoforms.

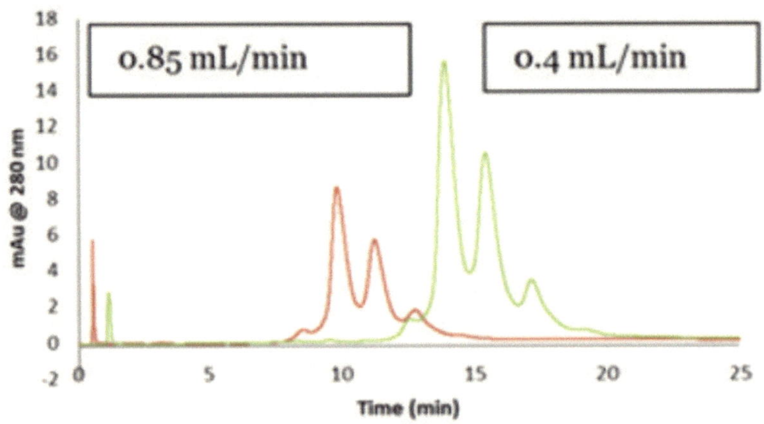

Figure 16.
Increased sensitivity at lower flow rate.

6.2 Gradient slope

The gradient elution method is common for the separation of species of different binding strength. **Figure 17** shows the effect of a gradient slope on the separation mAb on TSKgel FcR-IIIA-NPR. As expected, longer gradient time increased the resolution between the peaks whereas the overall peak area and relative ratio of the peak areas remained unchanged. **Figure 17** indicates how shallower slope increased the resolution between the three peaks. This is particularly noticeable between peaks 2 and 3. On the other hand, it should be noted that, for any further analytical

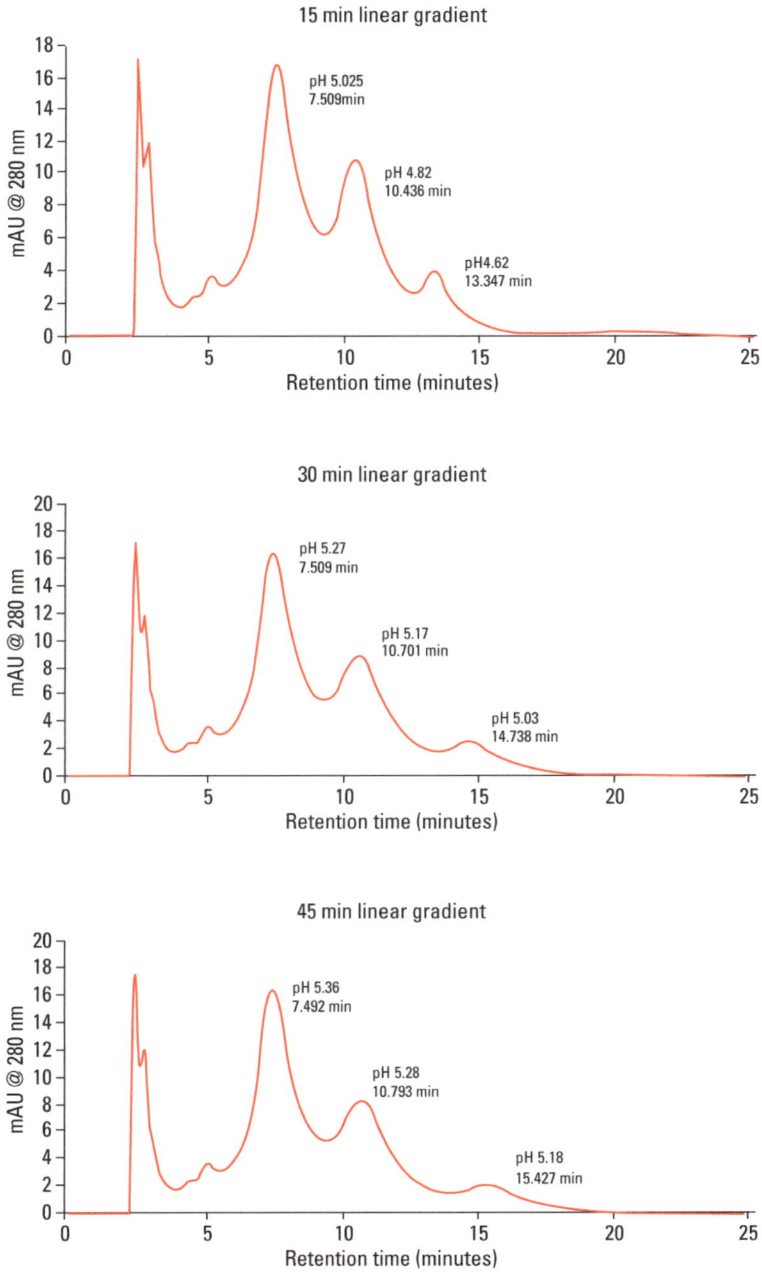

Figure 17.
Effect of gradient slope on the separation efficiency of peaks.

work (e.g. for mass spec), longer gradient increases the peak volumes and thus peak fractions will be more diluted.

6.3 Temperature

The ligand in FcR-IIIA-NPR column is a 20-kDa folded polypeptide and thus special care is to be considered to maintain protein conformation intact with proper run temperature. **Figure 18** shows the separation of mAb at four different temperatures (5°C, 15°C, 20°C, 25°C) at flow rate of 0.2 mL/min. As the temperature increased, the retention time of the three peaks decreased, indicating somewhat lower binding affinity as a function of higher temperature. However, importantly, overall peak profile at each temperature remains unchanged. Thus, for practical reasons, temperature range from 15 to 25°C is recommended for most analytical work. Following the completion of the analysis the column needs to be stored at 2–8°C.

6.4 Sample load

Figure 19 shows the effect of load amount of mAb on the separation profile. The limit of detection was determined as 1.5 μg as per USP definition S/N of 2–3. A load of 3.16 μg could still be easily quantified (LOQ). The analysis was repeatable, robust and the total peak area increased proportionately as the load amount was

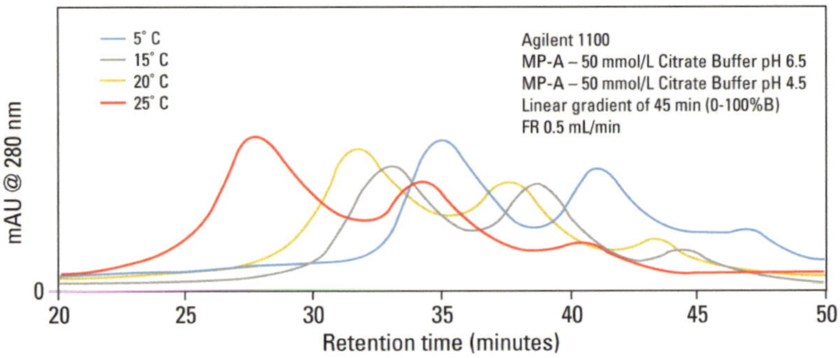

Figure 18.
Effect of temperature on the separation of mAb.

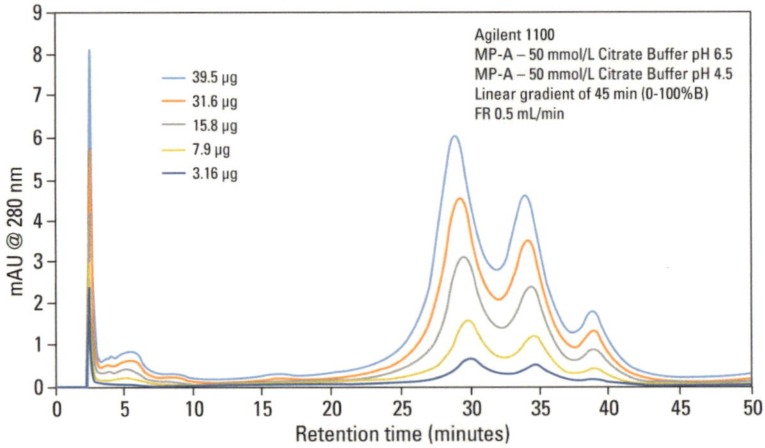

Figure 19.
Effect of sample load on the separation of mAb.

increased in a linear manner in consecutive injections. Relative ratio of the individual peak areas in the three peaks remained constant. The column can generally be used up to 100 μg protein load. However, 5–50 μg load of mAb is recommended for the best resolution and for maintaining the lifetime of the column.

Presence of aggregates in IgG samples impact the binding to the Fcγ receptors. A recently published article reports that deamidated IgG samples caused aggregation or formation of higher molecular weight (HMW) species and thereby impacted the binding affinity. Asparagine deamidation led to reduced binding of IgG to the low affinity FcγRIIIa receptor [24]. IgGs may also be more prone to aggregation when glycans are absent, which in turn has an effect on Fc effector functions. Lack of glycan and its effect on binding is explained below in Section 7.1. IgG dimers and aggregates may also bind stronger to different types of Fc receptors and thus have significant impact on affinity determination. Accumulated strength of multiple non-covalent affinities between the ligand and the receptor is known as avidity effect. This effect can alter the binding to the receptor and should be considered during the analysis mAb with dimer and higher order aggregates. The interaction, if any, needs to be evaluated in case-by-case basis. Up to 5% of aggregates in IgG samples changed the binding and kinetics to each of the Fc receptors [24]. Methionine is easily oxidized to methionine sulfoxide which can also lead to light chain aggregation. Oxidation has impact on the binding to the Fcγ receptors and depends on the extent of oxidation. As reported [24], methionine oxidation below 7% did not impact on binding to the receptors. Taken together, all above factors should be considered when using this column, especially during analysis method development for mAbs containing any amount of aggregates or oxidized forms.

7. Robustness of TSKgel FcR-IIIA-NPR affinity column

7.1 Selectivity

The usefulness of any affinity chromatography column depends on several robustness factors. Here, selectivity is dependent on the nature of N-glycan. This is clearly demonstrated by analyzing enzymatically deglycosylated mAb. PNGase-F deglycosidase reacts between asparagine residue and the innermost N-acetyl glucosamine (GlcNAc) of the complex oligosaccharide or high mannose content. **Figure 20** shows that enzymatically deglycosylated NIST mAb does not bind to the column and thus elutes in void volume.

7.2 Lot-to-lot variability

Scope of quality control of the therapeutic antibodies is expanding rapidly due to the emergence of biosimilars, “biobetter” forms and numerous other kinds of biologics in the biotherapeutic market. Lot-to-lot difference in the activity of innovator mAb may vary up to 20% in the manufacturing process [25]. Although substantial improvement has been attained in CHO cell engineering during recent years, and different strategies are there e.g., to produce afucosylated antibody drugs, still not enough technology is available to fully control *in vivo* glycosylation during production [26]. The lot-to-lot difference in N-glycan content may give rise to a wide variety of risk and thus N-glycan heterogeneity is a key factor to be monitored in quality control.

To demonstrate importance of the mAb lot-to-lot quality control, two manufacturing lots of mAb were analyzed using the TSKgel FcR-IIIA-NPR column (**Figure 21**). Both lots yielded a similar 3-peak elution profile. However, when percentual peak areas of the individual peaks were compared to check the consistency between the two mAb lots. Lot B showed a higher glycan percentage in peak

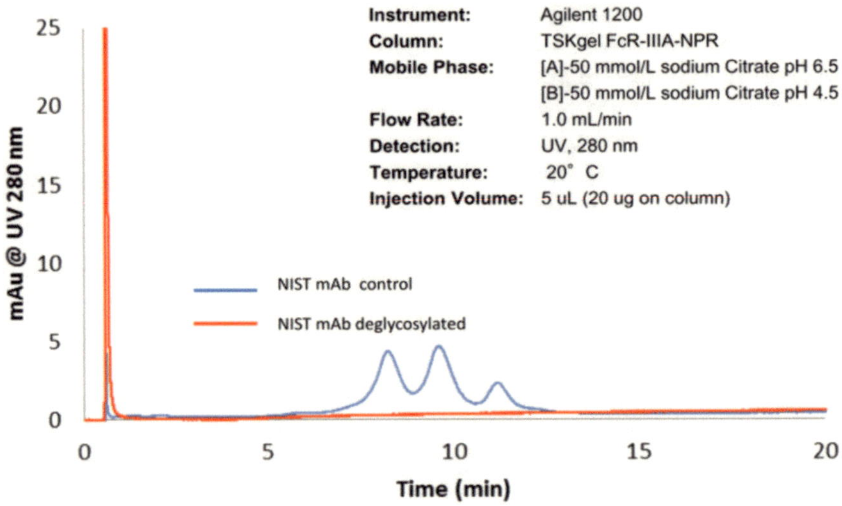

Figure 20.
Deglycosylated mAb does not bind to the column.

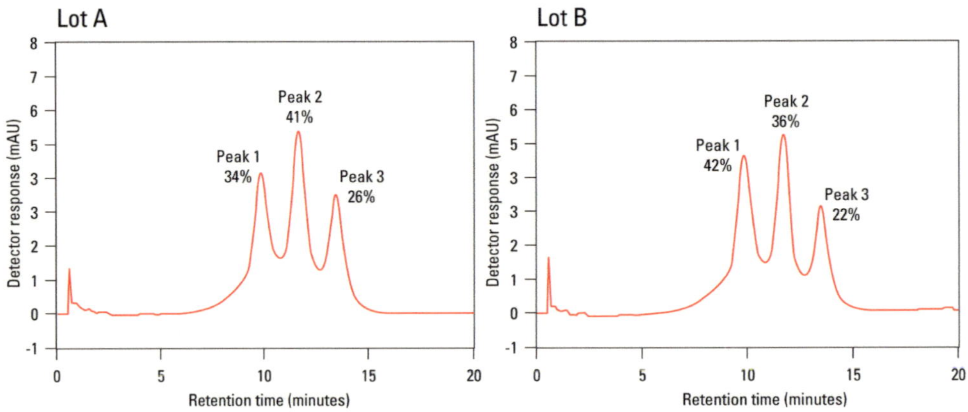

Figure 21.
N-glycan analysis of two manufacturing lots of a therapeutic antibody on TSKgel FcR-IIIA-NPR.

1 (42% versus 34%) and lower percentual amounts in peaks 2 and 3. In a subsequent ADCC assay, this also correlated with lower ADCC activity in the lot B. This experiment thus supports the notion that FcR affinity chromatography is suitable for lot-to-lot quality control of therapeutic mAbs.

To confirm consistency in FcR column manufacturing is also equally important for quality control. Three different lots of TSKgel FcR-IIIA-NPR column (Lots A, B, and C) were tested using reference mAb sample under identical chromatographic conditions (**Figure 22**). No significant variation in 3-peak profile was noticed between the three different column lots.

7.3 Effect of host cell proteins on the separation of mAb

Most mAb pharmaceuticals are produced in CHO cell culture system. Host cell proteins (HCPs), or host cell impurities, are collectively recognized as several forms of host cell products such as DNA, proteins, endotoxin and, if contaminated, viruses. These together are considered as process-related contaminants. They often have antigenic or pyrogenic effects in human and thus must be removed during downstream processing.

With regard to QC characterization, it is also necessary to assess if host cell proteins can interfere mAb binding on a TSKgel FcR-IIIA-NPR column. In the following study, CHO cell culture supernatant ("feedstock") was directly used for FcR column analysis and the results were compared to a previously purified mAb in the same assay (**Figure 23**, Panel a). However, no significant difference was noticed between the two profiles. This indicates that the HCPs had no significant impact on the mAb affinity to FcR column and about 5 μg of mAb in a feedstock was enough to obtain a suitable signal for monitoring process development in a bioreactor. The robustness of the assay was further tested using the mAb in CHO cell supernatant with 200 consecutive injections (**Figure 23**, Panel b). The total peak area remained constant with a % RSD (n = 10) of 0.79. It was also noticed that addition of NaCl to minimize unwanted non-specific interactions further improved durability at 20°C. The **Figure 23** (Panel c) shows how FcR column can be used for cell line selection and upstream monitoring. In this case, samples from CHO cell culture supernatant were collected, filtered, captured on protein A, and then injected to a TSKgel FcR-IIIA-NPR column. NaCl was added to improve separation. Glycoform changes in mAb were monitored over 14 days. The proportion of the intensities and peak areas of the three peaks significantly changed over the days that can be correlated to indicate changes in ADCC activity during

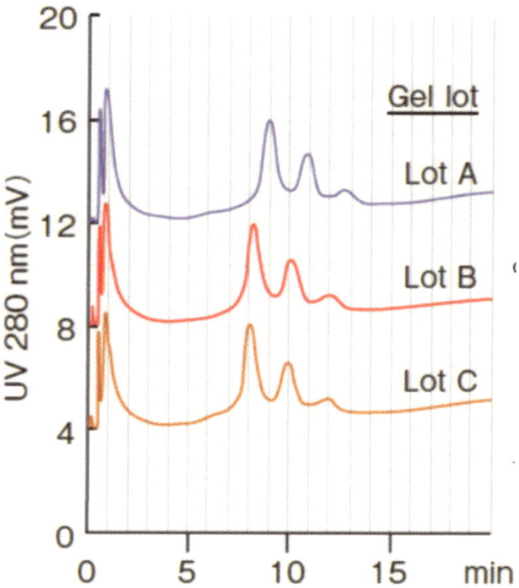

Figure 22.
Lot-to-lot consistency in TSKgel FcR-IIIA-NPR column manufacturing.

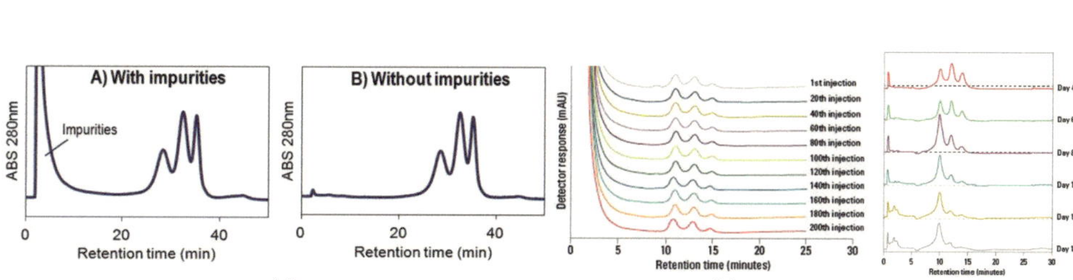

Figure 23.
(a) Analysis of CHO cell feedstock containing mAb versus purified mAb. (b) Assessment of FcR column stability over 200 injections. (c) Monitoring of glycan composition changes during fermentation.

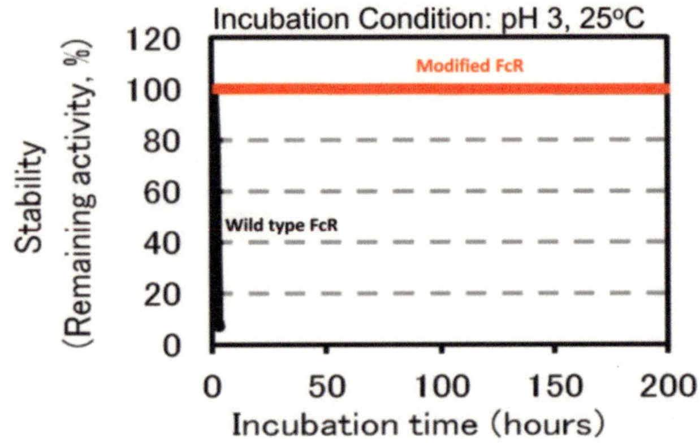

Figure 24.
Acid stability of a recombinant FcγRIII ligand as compared to a wild type ligand.

fermentation. Thus, by monitoring samples from the bioreactor using TSKgel FcR-IIIA-NPR, process engineers can approximate the optimal day for desired yield and ADCC activity.

7.4 Column pH stability and cleaning

Recommended working pH for the FcR column as mentioned in operational conditions and specifications (OCS) is from pH 8 to 4. As mentioned earlier, the protein ligand contains eight amino acid substitutions for improved stability. To further test acid stability, the column was held at pH3 for 200 hours. The modified ligand did not lose its binding affinity and the selectivity while the wild-type lost the binding affinity and selectivity within one hour (**Figure 24**). Based on this and other studies, a pH range of 3–8 is can be used for short term and pH 4.5–7 for long term usage. Due to a protein nature of the ligand, acetonitrile and other organic solvents are not suitable for the column. For cleaning, 3–5 injections of 0.5–2 mL of a buffer containing 500 mM NaCl or 20% ethanol can safely be used in reverse direction at half the normal flow rate. Once the cleaning procedure is complete, it is necessary to equilibrate the column in mobile phase for at least 45 minutes. Cleaning with alkalic solutions above pH 8 are not recommended since this will denature the protein ligand. Sodium azide (0.05%) can be used in the mobile phase as antibacterial agent.

8. Mass spectrometric characterization of glycoform peaks separated by TSKgel FcR-IIIA-NPR column

Mass spectrometric characterization is becoming an integral part of the liquid chromatography analysis. As an example how TSKgel FcR-IIIA-NPR column can be utilized in mass spec work, we describe here an in-line LC–MS intact mAb analysis of trastuzumab (Herceptin Biosimilar). The analysis was carried out using 100 mM volatile ammonium acetate buffer and a linear pH gradient from pH 6.5 to pH 4.5 at the flow rate of 0.4 mL/min. The wavelenght of detection was 280 nm. The column temperature during the analysis was maintained at 20°C. Three glycoform peaks could be detected by UV detector. Mass spectrometric detection was carried out using SCIEX X500B Q-TOF in ESI positive mode, within mass/charge (m/z) range of 5000–7000. Ion source gases 1 and 2 were maintained at 50 psi, curtain gas at 30 psi, CAD gas at 7 psi and temperature at 450°C. Spray voltage was maintained at 5200 V, declustering

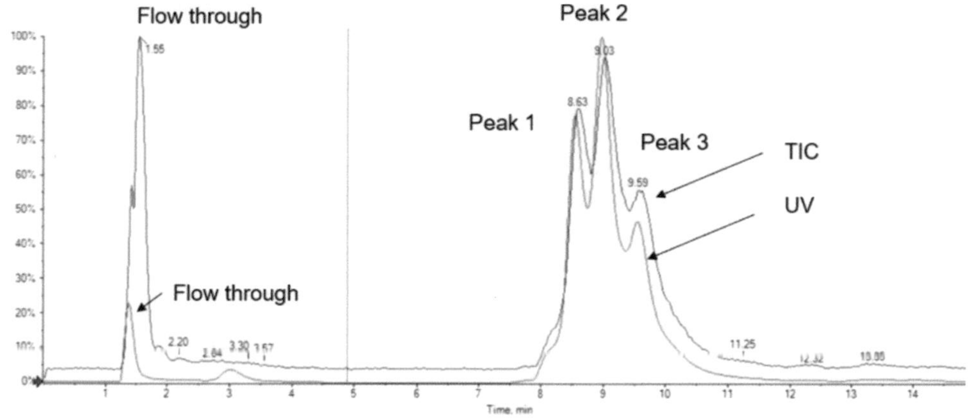

Figure 25.
Overlay of UV spectrum and Total Ion Chromatogram (TIC) of Herceptin Biosimilar.

potential at 275 V, and collison energy at 20 V. Time bins to sum was set at 120. For the automated characterization of the data acquired on the X500B QTOF, SCIEX Biotool kit software was used. Total Ion Chromatogram (TIC) was obtained by summing up intensities of all mass spectral peaks belonging to the same scan. An overlay of UV profile and TIC profile is shown in **Figure 25**. For further analysis of glyosylation profiles of these three peaks, SCIEX BioPharmaView™ software can be used.

As mentioned earlier, organic solvents such as acetonitrile are not suitable for the column and vapor pressure of water is very low. Thus, volatile salts such as 100 mM ammonium acetate or ammonium formate are used. To avoid ion source contamination during prolonged use, molarity should be kept at low (preferably <50 mM). Depending on the need for further optimization for different mAbs, volatile salt ammonium bicarbonate can also be used as such or in combination with other volatile salts.

9. Analysis of Fc fragment on TSKgel FcR-IIIA-NPR

The C-terminal part of the heavy chains contains the Fc fragment which is responsible for cellular effector functions, essential for proper function of most therapeutic mAbs. In some cases, it is desirable to express fragment antibodies that are smaller than intact mAbs but still are capable of eliciting their therapeutic function by activation of the immune system. Literature reports that both glycoengineering and protein engineering have rendered Fc domains with enhanced Fc receptor binding. In general, Fc fragments and their numerous variants are rapidly gaining interest as a platform in the development of efficient biotherapeutics.

The binding efficiency of the Fc fragment was tested in-house to assess suitability of TSKgel FcR-IIIA-NPR column on the characterization of smaller fragment antibodies. In short, trastuzumab was fragmented with papain that cleaves IgG at His228 forming two Fab parts and one Fc part (**Figure 26**). The reaction mixture was incubated at 37°C for 15 minutes to activate papain followed by mAb addition and further incubation overnight at room temperature. Papain activity was stopped with 5 mM iodoacetamide. A control a sample (no papain during incubation) and a sample from papain digestion were used for this study. Size exclusion chromatography followed by mass spectrometric analysis confirmed >95% cleavage of mAb to Fc and Fab fragments (data available by request).

Both the control mAb and fragments were analyzed on TSKgel FcR-IIIA-NPR column. As expected, Fab did not bind to the FcR column but eluted in

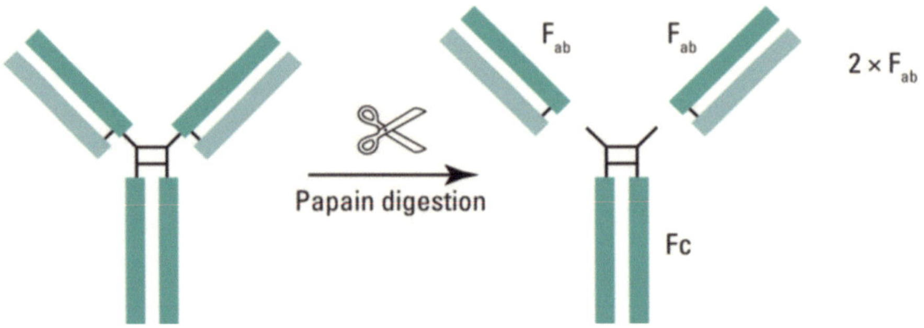

Figure 26.
Schematic representation of monoclonal antibody fragmentation with papain.

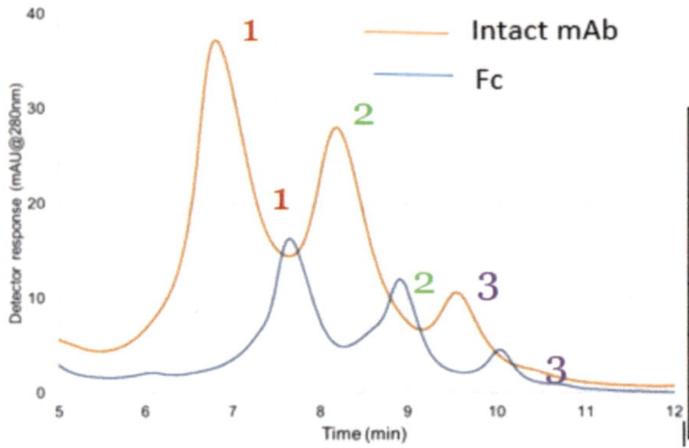

Peak #	pH at Peak Max (Calculated)	
	Intact mAb	Fc Fragment
1	5.09	5.00
2	4.90	4.84
3	4.73	4.65

Figure 27.
Analysis of intact mAb and Fc fragment on TSKgel FcR-IIIA-NPR column.

flow-through. Fc fragment efficiently bound to the column and yielded three glycoform peaks similar to intact mAb (**Figure 27**). Same sample volumes from the control sample and digestion reaction mixture were loaded onto the column. Lower peak heights for the Fc fragment were due to loss of Fab (2 x 48 kDa) from the protein mass during analysis. Interestingly, slightly longer retention times were detected for Fc fragment peaks, thus suggesting more rigid conformational stability for the Fc fragment leading to stronger binding as compared to the intact mAb. In summary, this experiment confirms that fragment antibodies, as long as they contain intact unobstructed Fc region, can be tested using the FcR column.

10. Novelty of TSKgel FcR-IIIA-NPR column

The mechanism of binding for IgG and other Fc engaging molecules is shown in the **Figure 28**. Complement component (C1q), Fc gamma receptors (FcγR), the Neonatal Fc receptor (FcRn), Tripartite motif 21 (Trim21), and Fc receptor-like (FcRL) molecules bind to various locations of mAb for the exertion of biological activity. For each ligand, the stoichiometric ratio of binding is also reported (Panel a). Recently, a biotinylated recombinant human FcRn immobilized to a Streptavidin Sepharose matrix and packed in a low pressure FPLC column has been introduced by Roche. A prepacked analytical protein-A affinity column (TSKgel ProteinA 5-PW) marketed by Tosoh also interacts with Fc region. Site of interaction of FcγR is separate from the site of interaction for FcRn or Protein A as seen in the **Figure 28** Panel b.

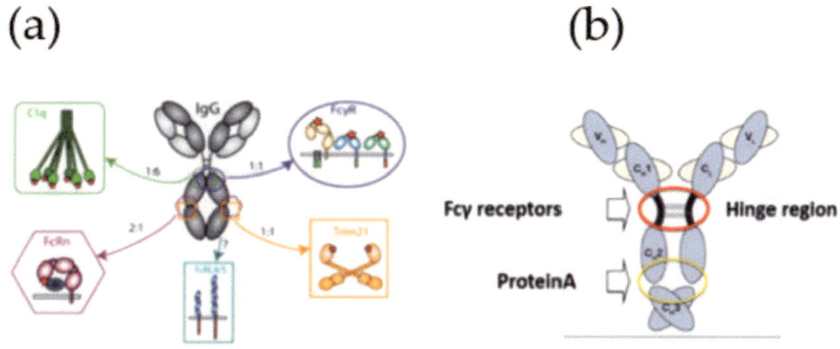

Figure 28.
Interaction of IgG with Fc effector molecules and protein A.

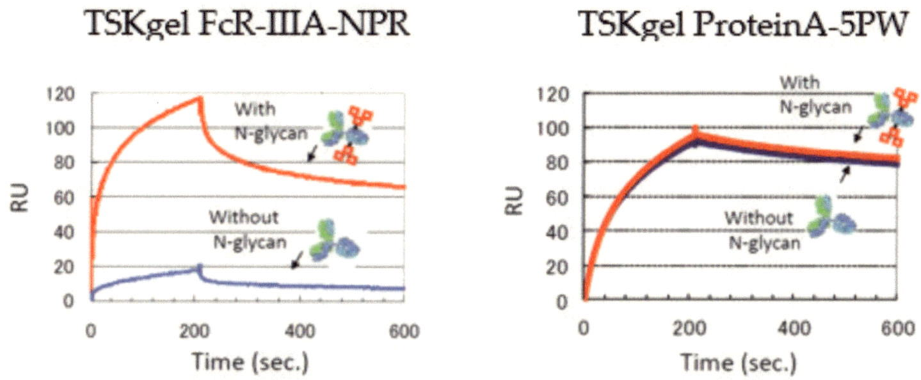

Figure 29.
Selectivity of the modified recombinant FcγRIIIa ligand vs. Protein A ligand.

TSKgel FcR-IIIA-NPR and TSKgel Protein A-5PW columns were tested for binding affinity of mAb with and without N-glycan using surface plasmon resonance technique (**Figure 29**). Protein A affinity chromatography column showed similar binding to mAb regardless of N-glycan whereas TSKgel FcR-IIIA-NPR column did not bind to mAb without N-glycan, similarly to the result in the **Figure 20**. Thus, the FcR column is unique due to its capability to analyze mAbs solely on the basis of their glycosylation.

11. Preparative scale purification of antibodies using FcR-based chromatography technique

Here, we provide a quick preview of the preparative scale TSKgel FcR-IIIA-5PW column that will be commercially available soon. The preparative column is manufactured using the same recombinant FcγRIIIA protein ligand, however, the ligand in this column is bound to porous (~100 nm nominal pore size) polymethacrylate polymer base beads. The column is suited for mAb purification in a significantly larger scale (loading 0.5–5 mg) as compared to the analytical FcR column (loading ≤100 μg). Chromatographic profile in the **Figure 30**, panel A was obtained using U/HPLC instrument and the analytical FcR column whereas the panel B shows peak separation with the preparative FcR column connected to a FPLC instrument. The peak separation profiles are closely similar with both columns. However, the preparative scale column allows collection of much more material for further experimentation such as glycan release, labeling and HILIC analysis, among other assays.

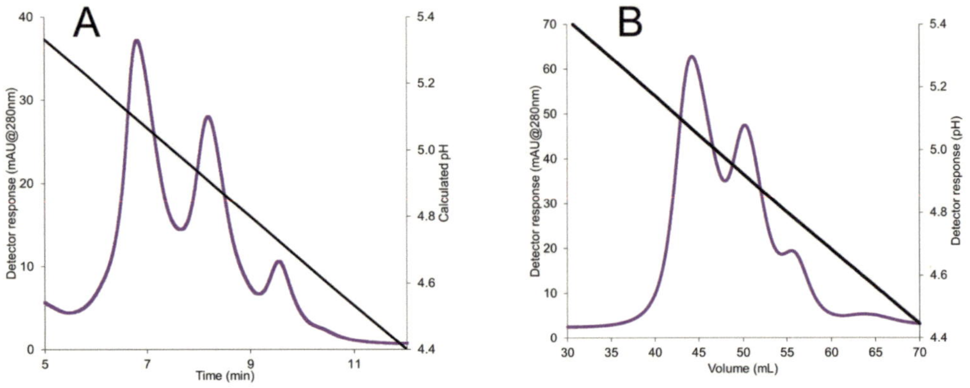

Figure 30.
Separation of mAb glycoforms using (A) analytical FcR column and (B) preparative FcR column.

12. Conclusions

TSKgel FcR-IIIA-NPR affinity chromatography column is a unique tool separating monoclonal antibodies into three peaks based on their glycosylation profile. Selectivity of the genetically engineered FcγRIIIa ligand is very specific to the mAb based on its glycan composition on highly conserved Asn-297 residue. IgG1, IgG3 and IgG4 subclasses bind to the FcR column whereas IgG2 subclass does not have affinity. IgA and IgM also don't bind to this column. Non-glycosylated mAb also does not bind to the column. Importantly, this column can be used for fast evaluation of antibody's ADCC effector function since the peak profile correlates well with the ADCC activity. Longer the retention time, higher is the ADCC activity.

The generally accepted workflow for mAb characterization, based on its glycan content typically follows the three different pathways (**Figure 31**, panel A). These are reporter bioassay for monitoring ADCC activity, Surface Plasmon Resonance (SPR) for measuring FcR affinity and U/HPLC-MS analysis for characterization of the glycan structure. Characterization of mAb on TSKgel FcR-IIIA-NPR column can combine these three pathways to one workflow (**Figure 31**, Panel B) in most circumstances.

The column is expected to be useful in several application areas (**Figure 32**) including (a) early screening of ADCC activity, (b) upstream (cell culture) optimization, (c) quality control of the mAb lot-to-lot consistency and (d) comparison between innovator and biosimilar products. Overall, this novel FcR affinity column is anticipated to be useful in research and development, characterization, manufacturing and quality control.

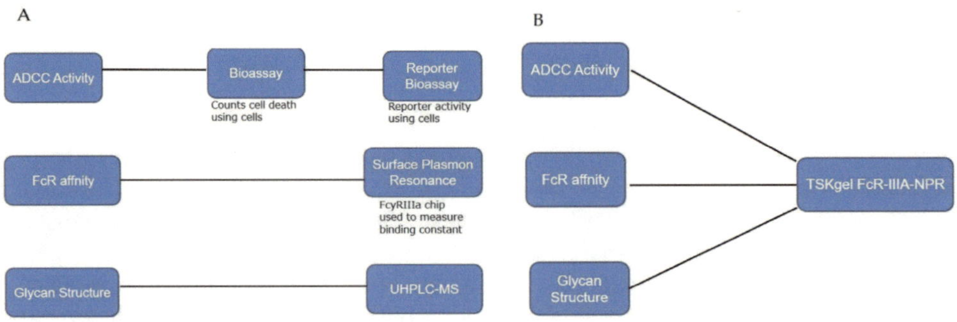

Figure 31.
Workflow for the characterization of mAb.

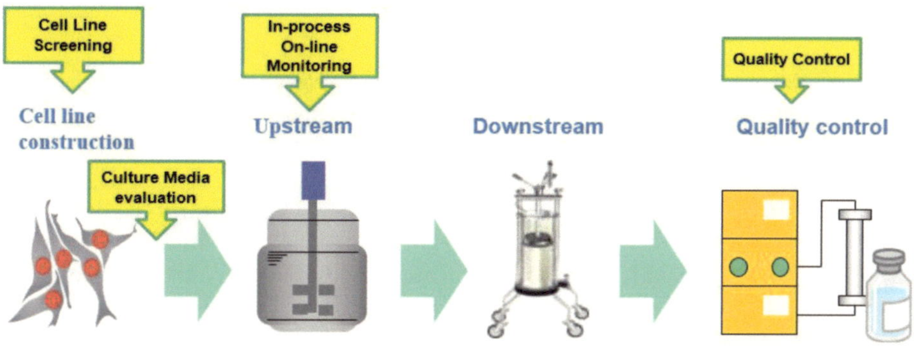

Figure 32.
Application areas of analytical and preparative FcR columns.

Acknowledgements

Chromatogram showing separation of mAb glycoforms using preparative TSKgel FcR-IIIA-5PW column (**Figure 30**, Panel B) was kindly provided by Scott L. Melideo, Ph.D., Tosoh Bioscience LLC. We sincerely thank Oscar Yamasaki and Hiroshi Tomizawa (Tosoh Bioscience, Tokyo, Japan) for their valuable comments and suggestions to the manuscript.

Conflict of interest

All the authors are current employees of Tosoh Bioscience, part of the Tosoh Corporation, that markets the TSKgel FcR-IIIA-NPR affinity column. Beyond this, the authors are not aware of any affiliations, memberships, funding, or financial holding that might be perceived as affecting the objectivity of this work.

Author details

Atis Chakrabarti[1*], Jukka Kervinen[1], Egbert Müller[2], Toru Tanaka[3] and Kazuaki Muranaka[3]

1 Tosoh Bioscience LLC, King of Prussia, PA, USA

2 Tosoh Bioscience GmbH, Darmstadt, Germany

3 Bioscience Division, Tosoh Corporation, Japan

*Address all correspondence to: atis.chakrabarti@tosoh.com

References

[1] Fanger MW, Graziano RF, Guyre PM. Production and use of anti-FcR bispecific antibodies. Immunomethods. 1994;4:72-81. DOI:10.1006/immu.1994.1009

[2] Mikulska J, Boratynski J, Niezgodka M, Lisowski, J. Human placental membrane receptor for IgG. Purification of the receptor and its subunit structure. Immunology Letters. 1982;5:137-143. DOI:10.1016/0165-2478(82)90098-0

[3] Vedeler CA, Matre R and Kristoffersen EK. Solubilization of human peripheral nerve Fc7 receptors and purification of a functional 40 kDa receptor. Immunology Letters. 1989;22:281-286. DOI:10.1016/0165-2478(89)90166-1

[4] Abuqayyas L, Balthasar JP. Investigation of the role of FcgammaR and FcRn in mAb distribution to the brain. Mol Pharm. 2013;10:1505-1513. DOI:10.1021/mp300214k

[5] Anderson CC, Sinclair NR. FcR-mediated inhibition of cell activation and other forms of coinhibition. Crit Rev Immunol. 1998;18:525-544. DOI:10.1615/critrevimmunol.v18.i6.30

[6] Bell CG. Characterization of the Fc receptors (FcR) functioning as bridge between antibody and cell-effector function; antibody-FcR-linked activation of immune responsiveness. Biochem Soc Trans. 1998;26:Article S200. DOI:10.1042/bst026s200

[7] Anania JC, Chenoweth AM, Wines BD, Hogarth PM. The Human FcγRI (CD32) Family of Leukocyte FcR in Health and Disease. Frontiers in Immunology. 2019; Article 464. DOI:10.3389/fimmu.2019.00464

[8] Wang H, Coligan JE, Morse HC 3rd. Frontiers in Immunology. Emerging Functions of Natural IgM and Its Fc Receptor FCMR in Immune Homeostasis. 2016;7:Article 99. DOI:10.3389/fimmu.2016.00099

[9] Kimberly RP, Salmon JE. and Edberg JC. Receptors for Immunoglobulin G. Arthritis and Rheumatism. 1995;38:306-314. DOI:10.1002/art.1780380303

[10] Lippold S, Nicolardi S, Wuhrer M and Falck D. Proteoform-Resolved FcγRIIIa Binding Assay for Fab Glycosylated Monoclonal Antibodies Achieved by Affinity Chromatography Mass Spectrometry of Fc Moieties. Frontiers in Chemistry. 2019; 7:Article 698. DOI:10.3389/fchem.2019.00698

[11] Dekkers G, Plomp R, Koeleman C.A.M, Visser R, von Horsten H.H, Sandig V, Rispens T, Wuhrer M and Vidarsson G. Multi-level glyco-engineering techniques to generate IgG with defined Fc-glycans. Nature. Scientific Reports. 2016;6:Article36964. DOI:10.1038/srep36964

[12] Indik ZK, Park J, Hunter S and Schreiber A.D. Structure/function relationships of Fey receptors in phagocytosis. Seminars in Immunology. 1995;7:45-54. DOI:10.1016/1044-5323(95)90007-1

[13] Radaev S, Motyka S, Fridman WH, Sautes-Fridman C, Sun PD. The structure of a human type III Fcgamma receptor in complex with Fc. The Journal of Biological Chemistry. 2001;276:16469-16477. DOI:10.1074/jbc.M100350200

[14] Bas M, Terrier A, Jacque E, Dehenne A, Pochet-Béghin V, Beghin C, Dezetter AS, Dupont G, Engrand A, Beaufils B, Mondon P, Fournier N, de Romeuf C, Jorieux S, Fontayne A, Mars LT, Monnet C. Fc Sialylation Prolongs Serum Half-Life of Therapeutic Antibodies. The Journal

of Immunology. 2019;202:1582-1594. DOI:10.4049/jimmunol.1800896

[15] Ferrara C, Grau S, Jäger C, Sondermann P, Brünker P, Waldhauer I, Hennig M, Ruf A, Rufer AC, Stihle M, Umaña P and Benz J. Unique carbohydrate–carbohydrate interactions are required for high affinity binding between FcγRIII and antibodies lacking core fucose. PNAS. 2011;108:12669-12674. DOI:10.1073/pnas.1108455108

[16] Liu Z, Gunasekaran K, Wang W, Razinkov V, Sekirov L, Leng E, Heather Sweet H, Foltz I, Howard M, Rousseau A, Kozlosky C, Fanslow W, and Yan W. Asymmetrical Fc Engineering Greatly Enhances Antibodydependent Cellular Cytotoxicity (ADCC) Effector Function and Stability of the Modified Antibodies. The journal of Biological Chemistry. 2014;289:3571-3590. DOI:10.1074/jbc.M113.513366

[17] Mizushima T, Yagi H, Takemoto E, Shibata-Koyama M, Isoda Y, Iida S, Masuda K, Satoh M and Kato K. Structural basis for improved efficacy of therapeutic antibodies on defucosylation of their Fc glycans. Genes to Cells. 2011;16:1071-1080. DOI:10.1111/j.1365-2443.2011.01552.x

[18] Wada R, Matsui M, Kawasaki N. Influence of N-glycosylation on effector functions and thermal stability of glycoengineered IgG1 monoclonal antibody with homogeneous glycoforms. MABS. 2019;11:350-372. DOI:10.1080/19420862.2018.1551044

[19] Kiyoshi M, Caaveiro J. M.M, Tada M, Tamura H, Tanaka T, Terao Y, Morante K, Harazono A, Hashii N, Shibata H, Kuroda D, Nagatoishi S, Oe S, Ide T, Tsumoto K, Ishii-Watabe A. Assessing the Heterogeneity of the Fc-Glycan of aTherapeutic Antibody Using an engineered FcgammaReceptor IIIa-Immobilized Column. Scientific reports. 2018;8:Article 3955. DOI:10.1038/s41598-018-22199-8

[20] Master Thesis (2019) of Leila Ghaleh, Technischen Universität Darmstadt, Rituximab kindly provided by Rentschler Biopharma in collaboration with Tosoh Bioscience GmbH

[21] Li T, DiLillo DJ, Bournazos S, Giddens JP, Ravetch JV, Wang LX. Modulating IgG effector function by Fc glycan engineering. Proc Natl Acad Sci USA. 2017;114:3485-3490. DOI:10.1073/pnas.1702173114

[22] de Taeye SW, Rispens T, Vidarsson G. The Ligands for Human IgG and Their Effector Functions Antibodies (Basel). 2019;8:2-18. DOI:10.3390/antib8020030

[23] Kang J, Kim SY, Vallejo D, Hageman TS, White DR, Benet A, Coghlan J, Sen KI, Ford M, Saveliev S, Tolbert TJ, Weis DD, Schwendeman SP, Ruotol BT, Schwendeman A. Multifaceted assessment of rituximab biosimilarity: The impact of glycan microheterogeneity on Fc function. European Journal of Pharmaceutics and Biopharmaceutics. 2020;146:111-124. DOI:10.1016/j.ejpb.2019.12.003

[24] Geuijen KPM, Oppers-Tiemissen C, Egging DF, Simons PJ, Louis Boon L, Schasfoort RBM and Eppink MHM. Rapid screening of IgG quality attributes – effects on Fc receptor binding. FEBS Open Bio. 2017;7:1557-1574. DOI:10.1002/2211-5463.12283

[25] Schiestl M, Stangler T, Torella C, Cepeljnik T, Toll H, Grau R. Acceptable changes in quality attributes of glycosylated biopharmaceuticals. Nature Biotech. 2011;29:310-312. DOI:10.1038/nbt.1839

[26] Pereira NA, Chan KF, Lin PC, and Song Z. The "less-is-more" in therapeutic antibodies: Afucosylated anti-cancer antibodies with enhanced antibody-dependent cellular cytotoxicity. MAbs. 2018;10:693-711. DOI:10.1080/19420862.2018.1466767

Chapter 5

In silico Techniques for Prospecting and Characterizing Monoclonal Antibodies

Tania M. Manieri, Carolina G. Magalhaes, Daniela Y. Takata, João V. Batalha-Carvalho and Ana M. Moro

Abstract

In the past few years, improvement in computational approaches provided faster and less expensive outcomes on the identification, development, and optimization of monoclonal antibodies (mAbs). *In silico* methods, such as homology modeling, to predict antibody structures, identification of epitope-paratope interactions, and molecular docking are useful to generate 3D structures of the antibody–antigen complexes. It helps identify the key residues involved in the antigen–antibody complex and enable modifications to enhance the antibody binding affinity. Recent advances in computational tools for redesigning antibodies are significant resources to improve antibody biophysical properties, such as binding affinity, solubility, stability, decreasing the timeframe and costs during antibody engineering. The immunobiological market grows continuously with new molecules, both natural and new molecular formats, such as bispecific antibodies, Fc-antibody fusion proteins, and mAb fragments, requiring novel methods for designing, screening, and analyzing. Algorithms and software set the *in silico* techniques on the innovation frontier.

Keywords: antibody structure modeling, VH-VL, computational analyses, epitope prediction, paratope prediction, molecular docking

1. Introduction

The development of new therapeutic antibodies is a multiple task challenge. The approval of OKT3 (1986), the first therapeutic mAb, opened the perspective of using this class of product in many other antibody-based therapies. Only the concept of "Magic Bullets," however, was not enough to provide for safety and efficacy, resulting in many preclinical or clinical trial failures. Soon became evident the need for the humanization of antibodies that mitigated their immunogenicity with the counterpart of decreasing their affinity in many cases. An alternative to circumvent this issue relied on back mutations. However, how to suggest such mutations?

Methodologies have emerged to optimize newly discovered antibodies, either in their affinity to the target or other properties, including *in silico* methods. The computational capacity has grown exponentially over the past few decades, providing an equally exponential advance in computational drug optimization techniques. The increased public's databases associated with bigdata works transformed the internet in the most profitable "laboratory" with free reagents (i.e., data), low-risk experiments (i.e., *in silico assays*), and time safe results. Considering all these aspects together are a simple way to understand the webtools' strengths.

In this chapter, we present different aspects of *in silico* methodologies for prospecting and characterizing mAbs. For didactic reasons, we could start from modeling molecules by homology (Section 2), followed by the prediction of epitopes and paratopes, affinity maturation, and molecular docking in Section 3, and finally, the improvement of biophysical and biological properties in Section 4. We aim to present free tools currently available, highlighting their features and applications, allowing the readers to find the most appropriate way to solve their problems. Albeit many tools and their applicability are shown, we call attention to their sequence of use and refinement that is inherent to the particular questions to be answered. We remind you that all the tools presented here were available online and free of charge to the academy until the closing of this edition.

2. Antibody structure modeling

The ability of the antibodies to recognize a diverse set of antigens is acquired by V(D)J recombination and affinity maturation. These two mechanisms contribute to a large number of possible unique sequences of the antibodies, around 10^{11-15} [1–3].

Protein structures are strongly related to specificity and function, and their knowledge is crucial to analyze the antibody. Although many crystal structures are available in the Protein DataBank [4], this number is small (around 6700) compared to the number of possible sequences. Computational modeling is a feasible method for predicting antibodies' structures and allows us to evaluate antibodies' properties and to understand antibody–antigen interaction.

The first step in the antibody modeling is its alignment with the germline sequence and the V(D)J classification. The International Immunogenetics Information System (IMGT) [5] is the central database of germline antibody sequences. Some webtools, such as IgBlast [6] and IMGT/V QUEST [7], use this database to align and classify the annotated sequence (**Table 1**). Since differences in the variable domains are responsible for the structural and functional antibodies' diversity, most of the structure prediction methods are based on the Fv modeling (**Table 1**). Framework regions are sequences with highly conserved structures, making it easier to generate their models from template structures. CDRs from the light chain (CDRs L1–3), CDRs 1 and 2 from the heavy chain (CDRs H1–2) are relatively conserved, regarding their structures, being possible to predict their structures based on their amino acid sequence. There is a set of canonical structures that allows us to predict the conformation of each loop. Recent studies have classified non-CDR H3 loops by their type and length and identified 72 clusters [62]. CDR H3 loop is usually longer (5–26 amino acids) than the others and presents a highly diverse structure. The CDR H3 loop also influences the VL-VH orientation and, consequently, the antibody–antigen interaction [63, 64]. For these reasons, the major challenge in antibody modeling is to achieve accuracy in CDR3 loop structure prediction. Usually, the primary sequence of CDR H3 is not enough for the prediction of the loop conformation.

Some information, as the position of the key residues, seems to be necessary. CDR H3 can be divided into two regions, the torso and the head [65]. The head usually presents a standard hairpin structure [65]. The torso, which is the region closer to the framework, can be predicted comparing with similar antibodies in which crystal structures are available on databases. Some software as Rosetta [66] has a platform to predict antibody structure, which first models each CDR

DOI: http://dx.doi.org/10.5772/intechopen.94366

Antibody structural modeling	**Link**	**Ref.**
IgBLAST	https://www.ncbi.nlm.nih.gov/igblast/	[6]
IMGT V-Quest	http://www.imgt.org/IMGTindex/V-QUEST.php	[7]
SPHINX	http://opig.stats.ox.ac.uk/webapps/newsabdab/sabpred/sphinx	[9]
AbodyBuilder	http://opig.stats.ox.ac.uk/webapps/sabdab-sabpred/Modelling.php	[10]
LYRA	http://www.cbs.dtu.dk/services/LYRA/index.php	[11]
Kotai Antibody Builder	https://sysimm.org/rep_builder/	[12]
Rosetta Suite	http://rosie.rosettacommons.org/	[13]
PEARS	http://opig.stats.ox.ac.uk/webapps/sabdab-sabpred/PEARS.php	[14]
SCWRL	http://dunbrack.fccc.edu/scwrl4/	[15]
BetaSCPWeb	http://voronoi.hanyang.ac.kr/betascpweb/	[16]
FREAD	http://opig.stats.ox.ac.uk/webapps/fread/php/	[17]
PLOP	http://www.jacobsonlab.org/plop_manual/plop_overview.htm	[18]
SCALOP	http://opig.stats.ox.ac.uk/webapps/sabdab-sabpred/SCALOP.php	[19]
Paratope prediction	**Link**	**Ref.**
Paratome	http://ofranservices.biu.ac.il/site/services/paratome/	[20]
Antibody i-Patch	http://opig.stats.ox.ac.uk/webapps/sabdab-sabpred/ABipatch.php	[21]
proABC-2	https://bianca.science.uu.nl/proabc2/	[22]
Parapred	https://github.com/eliberis/parapred	[23]
Epitope prediction	**Link**	**Ref.**
EpiPred	http://opig.stats.ox.ac.uk/webapps/sabdab-sabpred/EpiPred.php	[24]
PEASE	http://ofranservices.biu.ac.il/site/services/epitope/index.html	[25]
IEDB (database)	https://www.iedb.org/	[26]
Linear B-cells epitope	http://tools.iedb.org/bcell/	[27, 28]
DiscoTope	http://tools.iedb.org/discotope/	[29, 30]
Ellipro	http://tools.iedb.org/ellipro/	[31]
Epitome (database)	http://www.rostlab.org/services/epitome	[32]
Epitope Conservancy Analysis	http://tools.iedb.org/conservancy/	[33]
Epitope Cluster Analys	http://tools.iedb.org/cluster/	[34]
Antibody–antigen docking	**Link**	**Ref.**
ClusPro	https://cluspro.bu.edu/login.php	[35]
ZDOCK	http://zdock.umassmed.edu/	[36]
PatchDock	https://bioinfo3d.cs.tau.ac.il/PatchDock/	[37]
AutoDock	http://autodock.scripps.edu/	[38]
AutoDockFR	https://ccsb.scripps.edu/adfr/	[39]
FLIPDock	http://flipdock.scripps.edu/	[40]

Antibody structural modeling	**Link**	**Ref.**
Swarmdock	https://bmm.crick.ac.uk/~svc-bmm-swarmdock/	[41]
SnugDock	http://rosie.graylab.jhu.edu/snug_dock	[42]
HADDOCK	https://wenmr.science.uu.nl/haddock2.4/	[43]
FRODOCK	http://frodock.chaconlab.org/	[44]
Biophysical properties of mAbs	**Link**	**Ref.**
DeepDDG	http://protein.org.cn/ddg.html	[45]
ScooP	http://babylone.ulb.ac.be/SCooP/	[46]
Mupro1.0	http://sysbio.rnet.missouri.edu/multicom_toolbox/index.html	[47, 48]
Ease MM	https://sparks-lab.org/server/ease-mm/	[49]
Strum	https://zhanglab.ccmb.med.umich.edu/STRUM/	[50]
MCSM	http://biosig.unimelb.edu.au/mcsm/stability	[51]
SDM	http://marid.bioc.cam.ac.uk/sdm2/	[52]
TAP	http://opig.stats.ox.ac.uk/webapps/newsabdab/sabpred/tap	[53]
AgreeRATE-Disc	https://www.iitm.ac.in/bioinfo/aggrerate-disc/	[54]
ISMBLab-PPI	http://ismblab.genomics.sinica.edu.tw/predict.php?pred=PPI	[55]
Prediction of glycosylation spots	**Link**	**Ref.**
NetNGlyc 1.0	https://services.healthtech.dtu.dk/service.php?NetNGlyc-1.0	[56]
N-GlyDE	http://bioapp.iis.sinica.edu.tw/N-GlyDE/	[57]
GlycoMod	https://web.expasy.org/glycomod/	[58]
GlyConnect	https://glyconnect.expasy.org/home	—
GlycoSiteAlign	https://glycoproteome.expasy.org/glycositealign/	[59]
NetCGlyc	http://www.cbs.dtu.dk/services/NetCGlyc/	[60]
NetOGlyc	http://www.cbs.dtu.dk/services/NetOGlyc/	[61]

Table 1.
Free onlinetools and databases. Adapted from Norman et al., 2019 [8].

and framework based on very similar antibodies and then generates the CDR H3 conformations by assembling small peptides fragments. Software as SPHINX [9], uses *ab initio* modeling algorithm to predict CDR H3 conformation. Some software that performs the antibody structure modeling, and also the CDR H3 modeling, are listed in **Table 1**.

Antibody modeling is an essential step for most of the procedures discussed below, and the researcher must proceed according to the necessary refinement.

3. Antibody-antigen complex: methods of paratope and epitope prediction; molecular docking

In the past few years, improvement in computational approaches provided faster and less expensive outcomes on the identification, development, and optimization of monoclonal antibodies (mAbs). One of the leading goals of the rational development of antibodies is the identification of epitope-paratope interactions. 3D structures of the antibody–antigen complex using X-ray crystallography are the gold standard to reach the binding site information; however, these experimental methods can be money and time-consuming, and scarce to obtain. Thus, computational methods mean a rapid alternative across antibody discovery.

In silico methods, such as homology modeling, molecular docking, and interface prediction can be used to generate 3D models of the antibody–antigen complexes and to predict critical residues involved in antigen binding. Once the antibody–antigen contact residues are known, it can be computationally mutated to screen for residues that could increase antibody specificity and affinity against the target, if desired. Computational techniques to perform such a process fall into those that predict the paratope, the epitope, or the entire antibody–antigen complex (**Figure 1**).

3.1 Paratope prediction

Paratope represents the antibody amino acid residues in direct contact with the antigen. Since antigen-binding typically involves residues in the CDRs, about 80% of the amino acids constituting the paratope are in the CDRs [67]. However, only a third of the CDR residues participate in antigen binding [68]. Besides the residues in the CDRs, some framework regions are also involved in antigen-binding [67], relevant to identify the paratope residues precisely.

Several computational methods exist to predict paratopes (**Table 1**). The *online* tool Paratome [20] indicates the antigen-binding regions given the amino acid sequence or 3D structure. It identifies structural elements consensus, which is commonly involved in antigen binding between antibodies by aligning a set of all

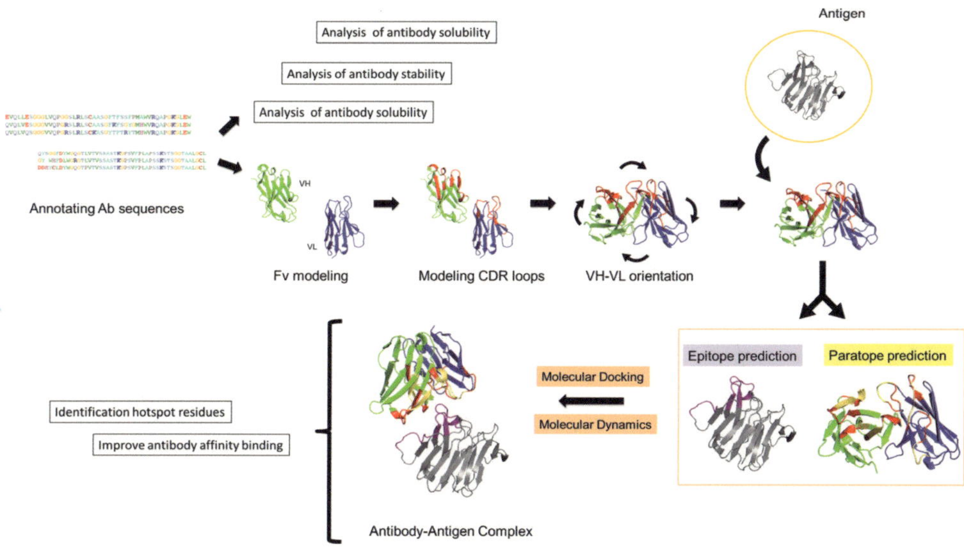

Figure 1.
Schematic representation of the computational approaches flows used in the antibody design and characterization. Methods to in silico *antibody analyses can be summarized into annotating the sequence, modeling the framework, modeling the CDR loops, and optimizing the VH-VL orientation, followed by predicting the paratope, the epitope, and the antibody–antigen contact residues. Mostly, the antibody design is performed, aiming the identification of hot spots residues suitable for mutagenesis to improve the antibody binding affinity to the target and antibody biophysical properties, such as stability, solubility. From the annotated sequence, it is possible to carry out the aggregation and solubility analyses and predict the glycosylation sites. These analyses can serve as verification steps throughout the process of antibody design and characterization.*

known antigen–antibody complexes on PDB. Paratome can also identify positions in the framework region that might contribute to antigen recognition [20, 67]. However, this tool does not provide information on the specific residues directly involved in the binding, relevant to antibody engineering experiments, such as *in silico* affinity maturation [69].

Statistical approaches such as Antibody i-Patch [21] utilizes structural information of both the antibody and antigen to generate its paratope prediction. This software assigns a score to each residue indicating how likely they are to be in contact with a given antigen. The higher score implies that the residue is more likely to form part of the paratope, information useful in guiding mutations in the artificial affinity maturation process. The tool considers the structure of both antibody and antigen. It should return more bespoke results, generating more accurate antigen-binding residue predictions.

Recently, machine learning approaches overcame Paratome and Antibody i-Patch. The PRediction Of AntiBody Contacts (proABC) [69] is a random forest algorithm, based on a machine-learning method, which uses the antibody sequences (eliminating the need for a 3D structure), the hypervariable loop canonical forms and lengths [70], and the germline family [71] as features to predict which residues of an antibody are involved in recognizing its cognate antigen. The prediction includes the nature of their contacts, distinguishing between hydrogen bonds, hydrophobic, and other non-bonded interactions. The proABC-2 [22] is an update of the original random-forest antibody paratope predictor, which uses the same set of features, but based on a deep learning framework, thus generating improved predictions and, as a consequence, increasing the success rate and quality of the docked models.

The Parapred [23] was the first algorithm based on modern deep learning for paratope prediction. This method only requires the amino acid sequence of a CDR and four adjacent residues as its input, without a full sequence, homology model, crystal structure, or antigen information. Its predictions improve the speed and accuracy of a rigid docking algorithm. The AG-Fast-Parapred [72] is an outperform of Parapred, which for the first time, provides antigen information in an in-depth paratope predictor.

Computational predictors of paratopes can provide valuable information to guide the modeling of antibody–antigen complexes. They will enable the accurate identification of residues that are the most important in determining the antibody's activity, leaving other residue positions as potential mutation sites, open to exploring other molecular characteristics by engineering.

3.2 Epitope prediction

The epitope is the antigen region in contact with the antibody in an antibody–antigen complex. Accurate identification of an epitope is a substantial step in characterizing the function of an antibody, helps predict possible cross-reactivity, and understand antibody mechanisms of action. The gold standard to determine the antibody epitope and the paratope is the 3D structure of the antibody–antigen complex by X-ray crystallography. Adding to these methods, peptide array, peptide ELISAs, phage display, expressed fragments, partial proteolysis, mass spectrometry, and mutagenesis analyses are also experimental methods applied to identify antibody epitopes. However, those assays can be expensive, time-consuming, and their outcome is uncertain [73].

Computational methods serve as an alternative to identify antibody epitopes (**Table 1**) [74]. Methods for computational B-cell epitope prediction can be categorized into the sequence and structure-based methods; the former focus on identifying contiguous stretches of primary amino acid sequence to predict linear epitopes. In contrast, the latter takes into account their 3D structure to predict conformational epitopes. The first *in silico* B-cell epitope prediction methods focused on amino acid properties within a sequence, such as hydrophobicity, hydrophilicity, or antigenicity [75–79]. They aim to identify propensities and patterns of a set of residues on the antigen capable of binding to an antibody [29]. However, in many epitope prediction methods, it lacks information on the cognate antibody resulting in limited practical use since the epitopes predicted are generical. Therefore, antibody-specific epitope prediction methods later replaced these approaches.

The first antibody-specific epitope prediction method was suggested in 2007 by Rapberger et al. [80]. Some approaches, such as ASEP [81], BEPAR [82], ABEpar [83], and PEASE [25], are antibody-specific epitope prediction methods that do not require antibody structure. The PEASE (Predicting Antibody-Specific Epitopes) method is based on a machine-learning model and utilizes the sequence of the antibody, in the absence of structural information. It evaluates a pair score for all combinations of residues from the antibody CDR and residues from the surface-exposed region of antigen. The predictions are provided both at the residue level and as patches on the antigen structure using antibody–antigen contact preferences and other properties computed from the antibody sequence and antigen structure or sequence. The EpiPred [24] is an antibody-specific epitope prediction method that identifies the epitope region on the antigen combining conformational matching of the antibody–antigen structures and a specific antibody–antigen score. Patches on the antigen structure are ranked according to how likely they are to be the epitope. This method aims to generate epitope predictions specific for a given antibody to facilitate docking.

The most recent approaches, such as MabTope [84] and the method suggested by Jespersen et al. [85], are docking-based prediction methods of the epitope. The MabTope methodology integrates both a docking-based prediction method and experiment steps. MabTope involves three phases; in the first, docking the antibody on its target to generate possible conformations of the antigen–antibody complex (docking poses); secondly, ranking these docking poses with the design of the peptides predicted to be part of the epitope; and last, experimental validation procedures based on these peptides. The method suggested by Jespersen et al. combines geometric and physicochemical features correlated in paratope-epitope interactions with statistical and machine learning algorithms. This method can identify the cognate antigen target for a given antibody, besides the antibody target for a given antigen.

Several B-cell epitope databases were developed over the last decades, compiling validated information of the experimentally annotated B-cell epitopes. The Immune Epitope Database (IEDB) [26] is a multifaceted database that includes epitope sequence and structure, source antigen, the organism from which the epitope is derived, and details of the experiments describing recognition of the epitope. IEDB provides tools to predict linear B-cell epitopes based on sequence characteristics of the antigen [27, 28], and also to predict B-cell epitopes from protein structure, using methods based on solvent-accessible surfaces, such as DiscoTope [29, 30] and ElliPro [31]. The database Epitome [32] compiles a collection of antibody–antigen complex structures, describes the residues (on antigen and antibody CDRs) involved in the interactions, and provides information concerning specific structural characteristics of the binding regions.

The epitope information from the B-cell epitope databases can evaluate existing epitope prediction methods and develop new and better algorithms for prediction. The identification or prediction of epitopes might be useful as an information for more sophisticated computational antibody design methods, such as antibody–antigen docking.

3.3 Antibody–antigen docking

The paratope and epitope prediction methods can offer useful information on antibody–antigen recognition by identifying a subset of residues involved in antigen–antibody interface formation. However, they do not provide information about the specific pairwise relations between the residues on the antibody and the antigen. This issue can be dealt with antibody–antigen docking, a specialized application of the broader field of molecular docking [86].

Molecular docking tools (**Table 1**) allow predicting the best binding interface of two interacting proteins. Different docking algorithms have been developed over the years to predict the 3D structure of biological complexes, and they typically involve two steps: sampling and scoring. In the sampling step, the conformational space surveys for thousands of possible complex conformations ('decoys'); in the scoring, the decoys are ranked using scoring functions, which sort the decoys to identify or predict the models that are closer to the native conformation (lowest energy structure). The sampling strategy applied during the simulation is used to classify the docking methods. The global docking algorithms do not consider any previous information about the binding interfaces and perform an exhaustive search of the interaction space. The local or integrative docking approaches, on the other hand, use the available experimental data or predicted information about the binding interface to drive the sampling during the docking [87].

There are three types of docking: rigid-body docking, partial flexible docking, and flexible docking [88, 89]. Most protein–protein docking algorithms perform rigid-body docking, which means that both binding partners are kept inflexible, as rigid molecules, hindering the exploration of conformational degrees of freedom during the binding. These methods are based on the fast Fourier transform search algorithm [90] and usually are applied when the structures are complementary [89]. Examples of used rigid-body docking software are ClusPro [35], ZDOCK [36], and PatchDock [37]. ClusPro is an antibody specific docking, unlike ZDOCK and PatchDock. In partial flexible docking, the antibody remains rigid, while the antigen is flexible [89]. One of the docking tools that applied this concept is AutoDock [38]. AutoDockFR [39] also allows partial flexibility of the antibody. However, removing the conformational limitations can improve the binding site identification, since, in most situations, protein flexibility is a crucial factor to be considered [91]. Therefore, flexible docking involves both interacting molecules as flexible structures. FLIPDock [40], Swarmdock [41], SnugDock [42], and HADDOCK [92–94] are examples of these approaches. SnugDock and HADDOCK allow some flexibility alongside chains and the backbone during a refinement stage. Snugdock is the first antibody specific docking to apply flexibility to the target antibody resulting in flexible binding interfaces, which can compensate for the errors caused by homology modeling [42].

The docking approaches depend on the 3D structures of the components. For antibodies, modeling methods can generate reasonably accurate structures [10, 95, 96]. Since these methods cannot compete with the reliability of crystallography-derived structures, the performances of docking methods are continuously evaluated by the Critical Assessment of Predicted Interactions (CAPRI) experiment [97, 98].

Although there are many successful cases in predicting the protein–protein complexes, docking of antibody–antigen complexes is still challenging [99–101] due to the inherent properties of their interfaces [102, 103]. As the improvement of predicting antibody–antigen interaction methods, we expect that the results of paratope prediction, epitope prediction, and antibody–antigen docking methods would offer a valuable, fast and economical alternative to obtain reliable information about which to base rational antibody design decisions (**Figure 1**).

3.4 *In silico* affinity maturation

Recent advances in computational prediction of the 3D structure of an antibody–antigen complex stimulated the development of *in silico* methods for redesigning antibodies to improve their biophysical properties, such as binding affinity. These computational methods can screen a large number of variants in a virtual library, in a short timeframe and a cost-effective manner, and select the one most optimized, based on a better understanding of antibody–antigen interactions and structural analysis through different algorithms.

The availability of crystal structures of antibody–antigen complexes is an essential factor in achieving computational antibody affinity maturation. However, when the crystal structures of the complexes are not available, as seen above, many modeling software can predict the 3D structure of the antibody–antigen complex [104]. When we use molecular docking for this purpose, it is possible to identify residues involved in intermolecular interactions and select candidate residues that can be mutated to improve antibody affinity [99, 102–105].

The prediction of binding affinities usually utilizes energy functions, such as physics-based force fields or knowledge-based statistical potentials derived from the structural database, to estimate changes in the free energy of an antibody–antigen complex with a focus on getting the global minimum energy conformation [106]. Some algorithms and methods identify the lowest energy function of two-body interactions through changes made in the amino acid sequence or the rotameric state of an amino acid [107, 108]. Computational tools, such as molecular dynamics, simulate the dynamic behavior of antibody structures, and provide alternative candidates that can be evaluated by further experimental assessments [89, 109]. Also, some tools can identify hotspot residues on protein interfaces, for which mutation to alanine strongly attenuates binding, and calculate the values for the change in the binding energy of the protein complex upon mutation [110–112]. These platforms are useful to study the effect of a particular amino acid on the binding affinity of an antibody–antigen complex.

Computational affinity maturation usually focuses on residues in the CDRs. However, as we learned in previous sections of this chapter, some residues in the framework can also play a role in the binding affinity and maintain the canonical conformations of antibodies. Although some mutations in noninteracting regions resulted in improved binding affinity [113, 114], the strategies to modify the CDR to increase antibody affinity are highlighted. Some examples of *in silico* affinity maturation of antibodies performed comprehensive computational CDR mutagenesis targeting all residues in CDRs or CDR H3 [115–118]. There are also examples of monitoring all interactions between the 3D structure of an antibody and its cognate target to determine the most relevant CDR residues in the binding by considering their stabilizing energies, inter and intra molecules distance, bonds formation or breakage, and overall complex stability [119].

These techniques still present deviations from the experimental data; however, they demonstrate that in some scenarios, computational approaches alone can be used for affinity maturation, decreasing the timeframe and costs of antibody engineering.

4. Analyses of mAbs' properties (solubility, stability, aggregation, chemical degradation, glycosylation)

In vitro antibody affinity maturation frequently results in a destabilizing process, needing compensatory modifications for preserving the thermodynamic stability of mAbs [120]. Emerging *in silico* tools are significant resources to promote the balance between affinity and stability during antibody engineering (**Table 1**). Before proceeding to available resources to deal with the destabilizing process, we should mention two types of stability in antibodies: physical and chemical. The physical stability of a protein is related to conformational changes and also to its colloidal stability. Concerning the conformational changes, we relate the free energy (ΔG) of the protein in its unfolded and folded-state, and the folded-state should present less energy than the unfolded state ($G_{folded\text{-}state} < G_{unfolded\text{-}state}$) [121]. One of the *in silico* methods used to investigate folded and unfolded-state energies was mentioned earlier, e.g., molecular dynamics.

Among the numerous *in silico* tools for predicting conformational stability, DeepDDG [45] proved to be quite efficient compared to eight other methods (**Table 1**). DeepDGG is a machine learning method trained from 5444 experimental data. This tool allows the calculation of the energy difference between the mutated protein and its native state. This calculation allows us to observe whether the proposed mutations, for example, for an improvement in affinity, cause structure destabilization. Experimentally, the conformational change of a protein it is accessed indirectly through its melting temperature (Tm), and it can be measured by different experimental techniques, such as scanning calorimetry (DSC), differential scanning fluorometry (DSF), and circular dichroism (CD). The changes between folded and unfolded-state can be reversible, unlike the process known as aggregation, related to colloidal stability.

Although aggregation is different from solubility, the solubility of a molecule is usually calculated for aggregation prediction. In computational chemistry, aggregation and solubility are commonly treated as the same parameter. The aggregation tendency of some mAbs that could impair their efficacy might be prevented through aggregation-prone regions (APRs) analyses. APR assays rely on the hydrophobicity scales and residues' charge annotations. Among several predictors of solubility and APRs for proteins, it is possible to highlight two endeavors successfully applied to antibodies: Wang et al. [122] combined tools to predict APRs in commercial mAbs. They found similar aggregation-prone motifs among commercial and non-commercial antibodies, without correlation with 3D structures.

In 2011, Agrawal et al. [123] compared several aggregation prediction tools demonstrating their usefulness in drug discovery and development, especially when screening a large number of molecules by fast and low cost *in silico* assays. Recently, Raybould et al. [53] launched Therapeutic Antibody Profiler (TAP), a web application that compares candidates' sequences with natural antibody sequences, as natural antibodies are assumed to display favorable biophysical properties. TAP, notably, depends on the previous data of clinical-stage antibody therapeutics (CST). So, the robustness of this method is directly affected by the input improvement of the CST database. One modern and elegant approach drove the development of AggreRATE-Disc [54], a machine learning-based tool that can predict, within the sequences, mutations that can promote or mitigate aggregation. Although *in silico* tools can highlight sequences with aggregation issues, they do not substitute experimental assays; however, they can be managed, reducing the totality of necessary tests. These tools and databases help the screening steps across the development/discovery of new therapeutic drugs.

Regarding the chemical stability of antibodies, it is possible to mention the degradation by chemical modification of amino acids, such as asparagine (Asn) deamidation, aspartate (Asp) isomerization, methionine (Met) oxidation, and lysine (Lys) glycation [124, 125]. The IgGs are commonly N-glycosylated at Asp297 residue in each Fc-CH_2 domain [126]. These Fc N-glycan are associated with correct folding, stability, aggregation, immunogenicity, and serum half-life of the mAbs. The conformational changes at the CH_2 antibody portion by multiple hydrophobic and polar non-covalent interactions harnesses the Fc binding to preferences of binding to C1q and FcɣRs [126]. There are no specific mAb glycosylation's webtools. Still, some web platforms (**Table 1**) designed to predict glycosylation sites on human protein sequences could also be useful for mAbs. The IgGs have a conservative N-glycan site; consequently, it needs attention in the engineering process that could accidentally create or remove a glycosylation site and interfering in the mAb chemical stability. In other instances, the glycosylation site is intentionally removed.

To evaluate any possible glycosylation spots, the NetNGlyc 1.0 [56] predicts N-glycosylation sites in human proteins using a trained neural network to distinguish between the acceptor and non-acceptor residue sequences. The N-GlyDE is a two-stage N-glycan prediction tool trained by the human proteome datasets. An algorithm generates a score between N-glycosylation proteins and non-N-linked glycoproteins in the first step. In the second stage, the prediction uses a support vector machine to evaluate if each asparagine-Xaa-serine/threonine (being Xaa different to proline) sequence can be glycosylated [57]. Further, the GlycoSiteAlign [59] is a tool that aligns amino acid sequences regarding its glycosylation site using the GlyConnect databank. This tool can be useful to compare a high number of mAbs sequences derived from different clones or expression conditions.

In a linear amino acid sequence of an antibody, it is possible to find numerous regions prone to modification. However, one must note that many of these regions may be buried due to the molecule conformation. Therefore, a conformational study is essential to highlight the residues liable to the chemical change. Chemical stability is generally based on statistical analysis derived from experiments or databases available in the literature, although some computational methods are being used [124, 127–133]. Statistics-based methods depend on data from previous experiments and provide valuable information about the behavior of proteins, being excellent guides during the development of new antibodies.

Currently, there are tools to predict the most varied protein characteristics. Many of them are free for academic purposes (**Table 1**). A difficulty still faced during the development of an antibody lies in the complexity of details and how one parameter influences another. For example, modifications to improve binding affinity may interfere with the stability of the molecule or even generate/remove a glycosylation spot. In the same way, a structural change for stability can impair binding affinity. There has been an immeasurable evolution of *in silico* methods, allowing analyzes to be carried out more quickly and at a lower cost than traditional experimental methods.

5. Highlights

Advances in bioinformatics allow us to outline different strategies in the discovery of new therapeutic antibodies. There has been significant progress in online tools in recent years, and probably the refinement of the techniques will be increased, bringing more accurate and reliable results.

Online platforms can present a long wait and execution times. The use of those platforms requires a good internet connection, and also a robust computer for analysis and treatment of the generated data.

Bioinformatics is a notably promising field, and indeed, has a prominent place on the innovation frontier.

Acknowledgments

FAPESP (2015/15611-0, 2016/08782-6, 2019/10724-2), CNPq (307636/2016-0). ORCID Tania M. Manieri (0000-0003-1152-7425), ORCID Carolina G. Magalhães (0000-0001-7099-060X), ORCID Daniela Y. Takata (0000-0001-6369-1775), ORCID João V. Batalha-Carvalho (0000-0002-1526-6915), ORCID Ana M. Moro (0000-0002-0650-7764).

Author details

Tania M. Manieri[1*], Carolina G. Magalhaes[1], Daniela Y. Takata[1], João V. Batalha-Carvalho[1] and Ana M. Moro[1,2]

1 Biopharmaceuticals Lab, Butantan Institute, Sao Paulo, Brazil

2 iii/INCT, São Paulo, Brazil

*Address all correspondence to: tania.manieri@butantan.gov.br

References

[1] Rees AR. Understanding the human antibody repertoire. MAbs. 2020;12(1):1729683.

[2] Boyd SD, Joshi SA. High-Throughput DNA Sequencing Analysis of Antibody Repertoires. Microbiol Spectr. 2014;2(5).

[3] Schroeder HW, Jr. Similarity and divergence in the development and expression of the mouse and human antibody repertoires. Dev Comp Immunol. 2006;30(1-2):119-35.

[4] Berman HM, Westbrook J, Feng Z, Gilliland G, Bhat TN, Weissig H, et al. The Protein Data Bank. Nucleic Acids Res. 2000;28(1):235-42.

[5] Lefranc MP, Pommie C, Kaas Q, Duprat E, Bosc N, Guiraudou D, et al. IMGT unique numbering for immunoglobulin and T cell receptor constant domains and Ig superfamily C-like domains. Developmental and comparative immunology. 2005;29(3):185-203.

[6] Ye J, Ma N, Madden TL, Ostell JM. IgBLAST: an immunoglobulin variable domain sequence analysis tool. Nucleic Acids Res. 2013;41(Web Server issue):W34-40.

[7] Brochet X, Lefranc MP, Giudicelli V. IMGT/V-QUEST: the highly customized and integrated system for IG and TR standardized V-J and V-D-J sequence analysis. Nucleic Acids Res. 2008;36(Web Server issue):W503-8.

[8] Norman RA, Ambrosetti F, Bonvin A, Colwell LJ, Kelm S, Kumar S, et al. Computational approaches to therapeutic antibody design: established methods and emerging trends. Brief Bioinform. 2019.

[9] Marks C, Nowak J, Klostermann S, Georges G, Dunbar J, Shi J, et al. Sphinx: merging knowledge-based and ab initio approaches to improve protein loop prediction. Bioinformatics. 2017;33(9):1346-53.

[10] Leem J, Dunbar J, Georges G, Shi J, Deane CM. ABodyBuilder: Automated antibody structure prediction with data–driven accuracy estimation. mAbs. 2016;8:1259-68.

[11] Klausen MS, Anderson MV, Jespersen MC, Nielsen M, Marcatili P. *LYRA*, a webserver for lymphocyte receptor structural modeling. Nucleic Acids Res. 2015;43(W1):W349-55.

[12] Schritt D, Li SL, Rozewicki J, Katoh K, Yamashita K, Volkmuth W, et al. Repertoire Builder: high-throughput structural modeling of B and T cell receptors. Mol Syst Des Eng. 2019;4(4):761-8.

[13] Lyskov S, Chou FC, Conchuir SO, Der BS, Drew K, Kuroda D, et al. Serverification of molecular modeling applications: the Rosetta Online Server that Includes Everyone (ROSIE). PLoS One. 2013;8(5):e63906.

[14] Leem J, Georges G, Shi J, Deane CM. Antibody side chain conformations are position-dependent. Proteins. 2018;86(4):383-92.

[15] Krivov GG, Shapovalov MV, Dunbrack RL, Jr. Improved prediction of protein side-chain conformations with SCWRL4. Proteins. 2009;77(4):778-95.

[16] Ryu J, Lee M, Cha J, Laskowski RA, Ryu SE, Kim DS. BetaSCPWeb: side-chain prediction for protein structures using Voronoi diagrams and geometry prioritization. Nucleic Acids Research. 2016;44(W1):W416-W23.

[17] Choi Y, Deane CM. FREAD revisited: Accurate loop structure

prediction using a database search algorithm. Proteins-Structure Function and Bioinformatics. 2010;78(6):1431-40.

[18] Jacobson MP, Pincus DL, Rapp CS, Day TJF, Honig B, Shaw DE, et al. A hierarchical approach to all-atom protein loop prediction. Proteins-Structure Function and Bioinformatics. 2004;55(2):351-67.

[19] Wong WK, Georges G, Ros F, Kelm S, Lewis AP, Taddese B, et al. SCALOP: sequence-based antibody canonical loop structure annotation. Bioinformatics. 2019;35(10):1774-6.

[20] Kunik V, Ashkenazi S, Ofran Y. Paratome: An online tool for systematic identification of antigen-binding regions in antibodies based on sequence or structure. Nucleic Acids Research. 2012;40:521-4.

[21] Krawczyk K, Baker T, Shi J, Deane CM. Antibody i-Patch prediction of the antibody binding site improves rigid local antibody-antigen docking. Protein Engineering, Design and Selection. 2013;26:621-9.

[22] Ambrosetti F, Olsen TH, Olimpieri PP, Jiménez-García B, Milanetti E, Marcatilli P, et al. proABC-2: PRediction Of AntiBody Contacts v2 and its application to information-driven docking. Bioinformatics. 2020.

[23] Liberis E, Velickovic P, Sormanni P, Vendruscolo M, Lio P. Parapred: Antibody paratope prediction using convolutional and recurrent neural networks. Bioinformatics. 2018;34:2944-50.

[24] Krawczyk K, Liu X, Baker T, Shi J, Deane CM. Improving B-cell epitope prediction and its application to global antibody-antigen docking. Bioinformatics. 2014;30:2288-94.

[25] Sela-Culang I, Ashkenazi S, Peters B, Ofran Y. PEASE: Predicting B-cell epitopes utilizing antibody sequence. Bioinformatics. 2015;31:1313-5.

[26] Vita R, Mahajan S, Overton JA, Dhanda SK, Martini S, Cantrell JR, et al. The Immune Epitope Database (IEDB): 2018 update. Nucleic Acids Res. 2019;47(D1):D339-D43.

[27] Chou PY, Fasman GD. Prediction of the secondary structure of proteins from their amino acid sequence. Adv Enzymol Relat Areas Mol Biol. 1978;47:45-148.

[28] Jespersen MC, Peters B, Nielsen M, Marcatili P. BepiPred-2.0: improving sequence-based B-cell epitope prediction using conformational epitopes. Nucleic Acids Research. 2017;45(W1):W24-W9.

[29] Kringelum JV, Lundegaard C, Lund O, Nielsen M. Reliable B Cell Epitope Predictions: Impacts of Method Development and Improved Benchmarking. PLoS Computational Biology. 2012;8.

[30] Andersen PH, Nielsen M, Lund O. Prediction of residues in discontinuous B-cell epitopes using protein 3D structures. Protein Science. 2006;15:2558-67.

[31] Ponomarenko J, Bui HH, Li W, Fusseder N, Bourne PE, Sette A, et al. ElliPro: A new structure-based tool for the prediction of antibody epitopes. Bmc Bioinformatics. 2008;9:1-8.

[32] Schlessinger A, Ofran Y, Yachdav G, Rost B. Epitome: database of structure-inferred antigenic epitopes. Nucleic acids research. 2006;34.

[33] Bui HH, Sidney J, Li W, Fusseder N, Sette A. Development of an epitope conservancy analysis tool to facilitate the design of epitope-based diagnostics and vaccines. Bmc Bioinformatics. 2007;8.

[34] Dhanda SK, Vaughan K, Schulten V, Grifoni A, Weiskopf D, Sidney J, et al. Development of a novel clustering tool for linear peptide sequences. Immunology. 2018;155(3):331-45.

[35] Kozakov D, Hall DR, Xia B, Porter KA, Padhorny D, Yueh C, et al. The ClusPro web server for protein-protein docking. Physiology & behavior. 2017;12:255-78.

[36] Pierce BG, Wiehe K, Hwang H, Kim BH, Vreven T, Weng ZP. ZDOCK server: interactive docking prediction of protein-protein complexes and symmetric multimers. Bioinformatics. 2014;30(12):1771-3.

[37] Schneidman-Duhovny D, Inbar Y, Nussinov R, Wolfson HJ. PatchDock and SymmDock: Servers for rigid and symmetric docking. Nucleic Acids Research. 2005;33:363-7.

[38] Morris GM, Huey R, Lindstrom W, Sanner MF, Belew RK, Goodsell DS, et al. AutoDock4 and AutoDockTools4: Automated docking with selective receptor flexibility. J Comput Chem. 2009;30(16):2785-91.

[39] Ravindranath PA, Forli S, Goodsell DS, Olson AJ, Sanner MF. AutoDockFR: Advances in Protein-Ligand Docking with Explicitly Specified Binding Site Flexibility. PLoS Comput Biol. 2015;11(12):e1004586.

[40] Zhao Y, Sanner MF. FLIPDock: docking flexible ligands into flexible receptors. Proteins. 2007;68(3):726-37.

[41] Torchala M, Moal IH, Chaleil RAG, Fernandez-Recio J, Bates PA. SwarmDock: A server for flexible protein-protein docking. Bioinformatics. 2013;29:807-9.

[42] Sircar A, Gray JJ. SnugDock: Paratope structural optimization during antibody-antigen docking compensates for errors in antibody homology models. PLoS Computational Biology. 2010;6.

[43] van Zundert GCP, Rodrigues J, Trellet M, Schmitz C, Kastritis PL, Karaca E, et al. The HADDOCK2.2 Web Server: User-Friendly Integrative Modeling of Biomolecular Complexes. J Mol Biol. 2016;428(4):720-5.

[44] Ramirez-Aportela E, Lopez-Blanco JR, Chacon P. FRODOCK 2.0: fast protein-protein docking server. Bioinformatics. 2016;32(15):2386-8.

[45] Cao H, Wang J, He L, Qi Y, Zhang JZ. DeepDDG: Predicting the Stability Change of Protein Point Mutations Using Neural Networks. Journal of chemical information and modeling. 2019;59(4):1508-14.

[46] Pucci F, Kwasigroch JM, Rooman M. SCooP: an accurate and fast predictor of protein stability curves as a function of temperature. Bioinformatics. 2017;33(21):3415-22.

[47] Cheng J, Randall AZ, Sweredoski MJ, Baldi P. SCRATCH: a protein structure and structural feature prediction server. Nucleic Acids Res. 2005;33(Web Server issue):W72-6.

[48] Cheng JL, Randall A, Baldi P. Prediction of protein stability changes for single-site mutations using support vector machines. Proteins-Structure Function and Bioinformatics. 2006;62(4):1125-32.

[49] Folkman L, Stantic B, Sattar A, Zhou Y. EASE-MM: Sequence-Based Prediction of Mutation-Induced Stability Changes with Feature-Based Multiple Models. J Mol Biol. 2016;428(6):1394-405.

[50] Quan L, Lv Q, Zhang Y. STRUM: structure-based prediction of protein stability changes upon single-point mutation. Bioinformatics. 2016;32(19):2936-46.

[51] Pires DE, Ascher DB, Blundell TL. mCSM: predicting the effects of mutations in proteins using

graph-based signatures. Bioinformatics. 2014;30(3):335-42.

[52] Pandurangan AP, Ochoa-Montano B, Ascher DB, Blundell TL. SDM: a server for predicting effects of mutations on protein stability. Nucleic Acids Res. 2017;45(W1):W229-W35.

[53] Raybould MIJ, Marks C, Krawczyk K, Taddese B, Nowak J, Lewis AP, et al. Five computational developability guidelines for therapeutic antibody profiling. Proc Natl Acad Sci U S A. 2019;116(10):4025-30.

[54] Rawat P, Kumar S, Michael Gromiha M. An in-silico method for identifying aggregation rate enhancer and mitigator mutations in proteins. Int J Biol Macromol. 2018;118 (Pt A):1157-67.

[55] Peng HP, Lee KH, Jian JW, Yang AS. Origins of specificity and affinity in antibody-protein interactions. Proc Natl Acad Sci U S A. 2014;111(26):E2656-65.

[56] Blom N, Sicheritz-Ponten T, Gupta R, Gammeltoft S, Brunak S. Prediction of post-translational glycosylation and phosphorylation of proteins from the amino acid sequence. Proteomics. 2004;4(6):1633-49.

[57] Pitti T, Chen CT, Lin HN, Choong WK, Hsu WL, Sung TY. N-GlyDE: a two-stage N-linked glycosylation site prediction incorporating gapped dipeptides and pattern-based encoding. Sci Rep. 2019;9(1):15975.

[58] Cooper CA, Gasteiger E, Packer NH. GlycoMod--a software tool for determining glycosylation compositions from mass spectrometric data. Proteomics. 2001;1(2):340-9.

[59] Gastaldello A, Alocci D, Baeriswyl JL, Mariethoz J, Lisacek F. GlycoSiteAlign: Glycosite Alignment Based on Glycan Structure. J Proteome Res. 2016;15(10):3916-28.

[60] Julenius K. NetCGlyc 1.0: prediction of mammalian C-mannosylation sites. Glycobiology. 2007;17(8):868-76.

[61] Steentoft C, Vakhrushev SY, Joshi HJ, Kong Y, Vester-Christensen MB, Schjoldager KT, et al. Precision mapping of the human O-GalNAc glycoproteome through SimpleCell technology. EMBO J. 2013;32(10):1478-88.

[62] North B, Lehmann A, Dunbrack RL, Jr. A new clustering of antibody CDR loop conformations. J Mol Biol. 2011;406(2):228-56.

[63] Noel D, Bernardi T, Navarro-Teulon I, Marin M, Martinetto JP, Ducancel F, et al. Analysis of the individual contributions of immunoglobulin heavy and light chains to the binding of antigen using cell transfection and plasmon resonance analysis. Journal of immunological methods. 1996;193(2):177-87.

[64] Xu JL, Davis MM. Diversity in the CDR3 region of V(H) is sufficient for most antibody specificities. Immunity. 2000;13(1):37-45.

[65] Morea V, Tramontano A, Rustici M, Chothia C, Lesk AM. Conformations of the third hypervariable region in the VH domain of immunoglobulins. J Mol Biol. 1998;275(2):269-94.

[66] Weitzner BD, Jeliazkov JR, Lyskov S, Marze N, Kuroda D, Frick R, et al. Modeling and docking of antibody structures with Rosetta. Nat Protoc. 2017;12(2):401-16.

[67] Kunik V, Peters B, Ofran Y. Structural consensus among antibodies defines the antigen binding site. PLoS Computational Biology. 2012;8.

[68] Esmaielbeiki R, Krawczyk K, Knapp B, Nebel JC, Deane CM. Progress and challenges in predicting protein interfaces. Briefings in Bioinformatics. 2016;17:117-31.

[69] Olimpieri PP, Chailyan A, Tramontano A, Marcatili P. Prediction of site-specific interactions in antibody-antigen complexes: The proABC method and server. Bioinformatics. 2013;29:2285-91.

[70] Chothia C, Lesk AM. Canonical structures for the hypervariable regions of T cell αβ receptors. J Mol Biol. 1987;196:901-17.

[71] Schatz DG, Swanson PC. V(D) J recombination: Mechanisms of initiation. Annual Review of Genetics. 2011;45:167-202.

[72] Deac A, Veličković P, Sormanni P. Attentive Cross-Modal Paratope Prediction. Journal of Computational Biology. 2019;26:536-45.

[73] Xu XL, Sun J, Liu Q, Wang XJ, Xu TL, Zhu RX, et al. Evaluation of spatial epitope computational tools based on experimentally-confirmed dataset for protein antigens. Chinese Science Bulletin. 2010;55:2169-74.

[74] El-Manzalawy Y, Honavar V. Recent advances in B-cell epitope prediction methods. Immunome Research. 2010;6:S2.

[75] Hopp TP, Woods KR. Prediction of protein antigenic determinants from amino acid sequences. Proceedings of the National Academy of Sciences of the United States of America. 1981;78:3824-8.

[76] Kolaskar AS, Tongaonkar PC. A semi-empirical method for prediction of antigenic determinants on protein antigens. FEBS Letters. 1990;276:172-4.

[77] Pellequer JL, Westhof E. PREDITOP: A program for antigenicity prediction. Journal of Molecular Graphics. 1993;11:204-10.

[78] Pellequer JL, Westhof E, Van Regenmortel MHV. Predicting location of continuous epitopes in proteins from their primary structures. Methods in Enzymology. 1991;203:176-201.

[79] Pellequer JL, Westhof E, Van Regenmortel MHV. Correlation between the location of antigenic sites and the prediction of turns in proteins. Immunology Letters. 1993;36:83-99.

[80] Rapberger R, Lukas A, Mayer B. Identification of discontinuous antigenic determinants on proteins based on shape complementarities. Journal of Molecular Recognition. 2007;20:113-21.

[81] Soga S, Kuroda D, Shirai H, Kobori M, Hirayama N. Use of amino acid composition to predict epitope residues of individual antibodies. Protein Engineering, Design and Selection. 2010;23:441-8.

[82] Zhao L, Li J. Mining for the antibody-antigen interacting associations that predict the B cell epitopes. BMC Structural Biology. 2010;10:1-13.

[83] Zhao L, Wong L, Li J. Antibody-specified B-cell epitope prediction in line with the principle of context-awareness. IEEE/ACM Transactions on Computational Biology and Bioinformatics. 2011;8:1483-94.

[84] Bourquard T, Musnier A, Puard V, Tahir S, Ayoub MA, Jullian Y, et al. MAbTope: A Method for Improved Epitope Mapping. The Journal of Immunology. 2018;201:3096-105.

[85] Jespersen MC, Mahajan S, Peters B, Nielsen M, Marcatili P.

Antibody specific B-cell epitope predictions: Leveraging information from antibody-antigen protein complexes. Frontiers in Immunology. 2019;10:1-10.

[86] Pagadala NS, Syed K, Tuszynski J. Software for molecular docking: a review. Biophysical Reviews. 2017;9:91-102.

[87] Ciemny M, Kurcinski M, Kamel K, Kolinski A, Alam N, Schueler-Furman O, et al. Protein–peptide docking: opportunities and challenges. Drug Discovery Today. 2018;23:1530-7.

[88] Guedes IA, de Magalhaes CS, Dardenne LE. Receptor-ligand molecular docking. Biophys Rev. 2014;6(1):75-87.

[89] Salmaso V, Moro S. Bridging molecular docking to molecular dynamics in exploring ligand-protein recognition process: An overview. Frontiers in Pharmacology. 2018;9:1-16.

[90] Katchalski-Katzir E, Shariv I, Eisenstein M, Friesem AA, Aflalo C, Vakser IA. Molecular surface recognition: Determination of geometric fit between proteins and their ligands by correlation techniques. Proceedings of the National Academy of Sciences of the United States of America. 1992;89:2195-9.

[91] Kotev M, Soliva R, Orozco M. Challenges of docking in large, flexible and promiscuous binding sites. Bioorg Med Chem. 2016;24(20):4961-9.

[92] de Vries SJ, van Dijk ADJ, Krzeminski M, van Dijk M, Thureau A, Hsu V, et al. HADDOCK versus HADDOCK: New features and performance of HADDOCK2.0 on the CAPRI targets. Proteins. 2007;69:726-33.

[93] Dominguez C, Boelens R, Bonvin AMJJ. HADDOCK: A protein-protein docking approach based on biochemical or biophysical information. Journal of the American Chemical Society. 2003;125:1731-7.

[94] van Zundert GCP, Melquiond ASJ, Bonvin AMJJ. Integrative Modeling of Biomolecular Complexes: HADDOCKing with Cryo-Electron Microscopy Data. Structure. 2015;23(5):949-60.

[95] Weitzner BD, Jeliazkov JR, Lyskov S, Marze N, Kuroda D, Frick R, et al. Modeling and docking of antibody structures with Rosetta. Nature Protocols. 2017;12:401-16.

[96] Yamashita K, Ikeda K, Amada K, Liang S, Tsuchiya Y, Nakamura H, et al. Kotai Antibody Builder: Automated high-resolution structural modeling of antibodies. Bioinformatics. 2014;30:3279-80.

[97] Janin J, Henrick K, Moult J, Eyck LT, Sternberg MJE, Vajda S, et al. CAPRI: A critical assessment of PRedicted interactions. Proteins: Structure, Function and Genetics. 2003;52:2-9.

[98] Méndez R, Leplae R, De Maria L, Wodak SJ. Assessment of blind predictions of protein-protein interactions: Current status of docking methods. Proteins: Structure, Function and Genetics. 2003;52:51-67.

[99] Pedotti M, Simonelli L, Livoti E, Varani L. Computational docking of antibody-antigen complexes, opportunities and pitfalls illustrated by influenza hemagglutinin. International Journal of Molecular Sciences. 2011;12:226-51.

[100] Ponomarenko JV, Bourne PE. Antibody-protein interactions: Benchmark datasets and prediction tools evaluation. BMC Structural Biology. 2007;7.

[101] Vajda S. Classification of protein complexes based on docking difficulty.

Proteins: Structure, Function and Genetics. 2005;60:176-80.

[102] Lo Conte L, Chothia C, Janin J. The atomic structure of protein-protein recognition sites. J Mol Biol. 1999;285:2177-98.

[103] Sela-Culang I, Kunik V, Ofran Y. The structural basis of antibody-antigen recognition. Frontiers in Immunology. 2013;4:1-13.

[104] Kuroda D, Shirai H, Jacobson MP, Nakamura H. Computer-aided antibody design. Protein Engineering, Design and Selection. 2012;25:507-21.

[105] Barderas R, Desmet J, Timmerman P, Meloen R, Casal JI. Affinity maturation of antibodies assisted by in silico modeling. Proceedings of the National Academy of Sciences of the United States of America. 2008;105:9029-34.

[106] Moal IH, Moretti R, Baker D, Fernández-Recio J. Scoring functions for protein-protein interactions. Current Opinion in Structural Biology. 2013;23:862-7.

[107] Desmet J, De Maeyer M, Hazes B, Lasters I. The dead-end elimination theorem and its use in protein side-chain positioning. Nature. 1992;356:539-42.

[108] Metropolis N, Rosenbluth AW, Rosenbluth MN, Teller AH, Teller E. Equation of state calculations by fast computing machines. The Journal of Chemical Physics. 1953;21:1087-92.

[109] González MA. Force fields and molecular dynamics simulations. Collection SFN. 2011;12:169-200.

[110] Schymkowitz J, Borg J, Stricher F, Nys R, Rousseau F, Serrano L. The FoldX web server: An online force field. Nucleic Acids Research. 2005;33:382-8.

[111] Tuncbag N, Keskin O, Gursoy A. HotPoint: Hot spot prediction server for protein interfaces. Nucleic Acids Research. 2010;38:402-6.

[112] Zhu X, Mitchell JC. KFC2: A knowledge-based hot spot prediction method based on interface solvation, atomic density, and plasticity features. Proteins: Structure, Function and Bioinformatics. 2011;79:2671-83.

[113] Fukunaga A, Tsumoto K. Improving the affinity of an antibody for its antigen via long-range electrostatic interactions. Protein Engineering, Design and Selection. 2013;26:773-80.

[114] Yanaka S, Moriwaki Y, Tsumoto K, Sugase K. Elucidation of potential sites for antibody engineering by fluctuation editing. Scientific Reports. 2017;7:1-10.

[115] Clark LA, Boriack-Sjodin PA, Eldredge J, Fitch C, Friedman B, Hanf KJM, et al. Affinity enhancement of an in vivo matured therapeutic antibody using structure-based computational design. Protein Science. 2006;15:949-60.

[116] Kiyoshi M, Caaveiro JMM, Miura E, Nagatoishi S, Nakakido M, Soga S, et al. Affinity improvement of a therapeutic antibody by structure-based computational design: Generation of electrostatic interactions in the transition state stabilizes the antibody-antigen complex. PLoS ONE. 2014;9:1-9.

[117] Lippow SM, Wittrup KD, Tidor B. Computational design of antibody-affinity improvement beyond in vivo maturation. Nat Biotechnol. 2007;25(10):1171-6.

[118] Willis JR, Sapparapu G, Murrell S, Julien JP, Singh V, King HG, et al. Redesigned HIV antibodies exhibit enhanced neutralizing potency and breadth. Journal of Clinical Investigation. 2015;125:2523-31.

[119] Ebrahimi Z, Asgari S, Ahangari Cohan R, Hosseinzadeh R,

Hosseinzadeh G, Arezumand R. Rational affinity enhancement of fragmented antibody by ligand-based affinity improvement approach. Biochemical and Biophysical Research Communications. 2018;506:653-9.

[120] Julian MC, Li L, Garde S, Wilen R, Tessier PM. Efficient affinity maturation of antibody variable domains requires co-selection of compensatory mutations to maintain thermodynamic stability. Sci Rep. 2017;7:45259.

[121] Kazlauskas R. Engineering more stable proteins. Chemical Society reviews. 2018;47(24):9026-45.

[122] Wang X, Das TK, Singh SK, Kumar S. Potential aggregation prone regions in biotherapeutics: A survey of commercial monoclonal antibodies. MAbs. 2009;1(3):254-67.

[123] Agrawal NJ, Kumar S, Wang X, Helk B, Singh SK, Trout BL. Aggregation in protein-based biotherapeutics: computational studies and tools to identify aggregation-prone regions. J Pharm Sci. 2011;100(12):5081-95.

[124] Kumar S, Plotnikov NV, Rouse JC, Singh SK. Biopharmaceutical Informatics: supporting biologic drug development via molecular modelling and informatics. The Journal of pharmacy and pharmacology. 2018;70(5):595-608.

[125] Lu X, Nobrega RP, Lynaugh H, Jain T, Barlow K, Boland T, et al. Deamidation and isomerization liability analysis of 131 clinical-stage antibodies. MAbs. 2019;11(1):45-57.

[126] Li W, Zhu ZY, Chen WZ, Feng Y, Dimitrov DS. Crystallizable Fragment Glycoengineering for Therapeutic Antibodies Development. Frontiers in Immunology. 2017;8.

[127] Yang R, Jain T, Lynaugh H, Nobrega RP, Lu X, Boland T, et al. Rapid assessment of oxidation via middle-down LCMS correlates with methionine side-chain solvent-accessible surface area for 121 clinical stage monoclonal antibodies. MAbs. 2017;9(4):646-53.

[128] Agrawal NJ, Dykstra A, Yang J, Yue H, Nguyen X, Kolvenbach C, et al. Prediction of the Hydrogen Peroxide-Induced Methionine Oxidation Propensity in Monoclonal Antibodies. J Pharm Sci. 2018;107(5):1282-9.

[129] Plotnikov NV, Singh SK, Rouse JC, Kumar S. Quantifying the Risks of Asparagine Deamidation and Aspartate Isomerization in Biopharmaceuticals by Computing Reaction Free-Energy Surfaces. J Phys Chem B. 2017;121(4):719-30.

[130] Jia L, Sun Y. Protein asparagine deamidation prediction based on structures with machine learning methods. PLoS One. 2017;12(7):e0181347.

[131] Delmar JA, Wang J, Choi SW, Martins JA, Mikhail JP. Machine Learning Enables Accurate Prediction of Asparagine Deamidation Probability and Rate. Mol Ther Methods Clin Dev. 2019;15:264-74.

[132] Sankar K, Hoi KH, Yin Y, Ramachandran P, Andersen N, Hilderbrand A, et al. Prediction of methionine oxidation risk in monoclonal antibodies using a machine learning method. MAbs. 2018;10(8):1281-90.

[133] Aledo JC, Canton FR, Veredas FJ. A machine learning approach for predicting methionine oxidation sites. Bmc Bioinformatics. 2017;18(1):430.

Chapter 6

Therapeutic Applications of Monoclonal Antibodies in Urologic-Oncology Management - An Update

Maya Kulshekar, Shridhar C. Ghagane, Sridevi I. Puranik, Rajendra B. Nerli and Murigendra B. Hiremath

Abstract

The idea of utilizing immunotherapy for the treatment of cancers has been appealing to scientists and clinicians for over a several decades. Immunotherapy for cancers encompasses knowledge gained from a wide range of disciplines and has the potential to procure the 'magic bullet' for the treatment of cancer. Monoclonal antibody-based treatment of cancer has been recognized as one of the most successful therapeutic strategies for both hematologic malignancies and solid tumours in the last 20 years. The discovery of hybridoma technology in late 1975 and the development of chimeric, humanized, and human antibodies have increased the availability and utility of immunotherapy for the treatment of cancer. Metastatic or recurrent cancer continues to be the bane of the urological oncologist. Despite recent improvements in therapeutic management and outcomes for clinically localized disease overall survival rate in patients with the majority of metastatic and recurrent genitourinary malignancies remains relatively unchanged. By targeting tumours through specific or associated antigens, it is possible to selectively eliminate tumour cells and maintain an acceptable toxicity profile. Therapeutic antibodies that target immune cells are also being developed with the goal of breaking local tolerance and stimulating the patient's anti-tumor immune response. As with other treatment modalities, immunotherapy is far from perfect and requires additional study to optimize clinical response and overcome therapeutic resistance. Modern advances in the field of immunotherapy hold the promise of providing the clinical urologist/oncologist with new tools to fight urological cancer. However, the literature on monoclonal antibody-based immunotherapy with a particular emphasis on target antigens, monoclonal antibody design and potential applications in the field of urology is limited. Hence, the present chapter focuses on the applications of Immunotherapy using monoclonal antibodies for urologic oncology settings such as prostate, bladder, renal, testicular and penile with a hope to highlight its clinical efficacy and also its mechanisms of action in each of these cancer types.

Keywords: Monoclonal antibodies, Immunotherapy, Applications, Urologic-oncology

1. Introduction

Modern advances and a quantum leap in the field of cancer therapy has been promising to oncologists with new tools to fight many cancers. The immune system has multifunctional units referred to as antibodies, mostly polyclonal which facilitate humoral and cellular reactions to antigens [1]. However, it is possible to produce large quantities of an antibody from a single B-cell clone which are called as *Monoclonal Antibodies (MAbs).* Using these antibodies for therapeutic purposes is termed as Immunotherapy. Immunotherapy in recent times has been propitious across a number of cancer types. Stimulating results with MAbs directed towards both established and emerging targets indicate its potential key role as a therapeutic agent [2]. These are being tested in early- and late-stage clinical trials. In the last 35 years over 100 Monoclonal Antibodies have been considered potential as drugs and many have been approved. The usage of the monoclonal antibodies in cancer therapy requires the understanding of the biological role of various antigens involved in tumor growth [3]. In cancer patients' immunity system is often altered. The purpose of immunotherapy with monoclonal antibodies is to interfere with synergic activity of immunosuppressive environment created by T cells, cytokines, interleukins and tumor growth factor [4]. In many cancer treatments, the monoclonal antibodies have been robust enough, however in some, combinatorial treatments including monoclonal antibodies, chemotherapy and vaccines have been successful thereby bringing together cancer immunologists and clinicians required for the management of cancer in the near future [5]. This chapter will focus on Immunotherapy using Monoclonal antibodies for many urologic oncology types such as prostate, renal, bladder, testicular and penile with a hope to highlight its clinical utility and also its mechanisms of action in each of these cancer types.

2. Types of monoclonal antibodies and their mode of action

There are several ways by which the mAbs are made. They are as follows:

- Human: Theses are derived from the human source. Called as 'umabs'
- Murine: These are derived from mouse. Called as 'omabs'
- Humanized: Here the mouse proteins are attached to the human protein. Called as 'zumabs'
- Chimeric: variable regions are from humans and constant regions are from mouse. Called as'ximabs'

Following are the types of Monoclonal antibodies:

1. **Naked monoclonal antibodies**

    a. These are the most common types of antibodies which are not attached to the radioactive material or any chemotherapy drugs. They act independently and have been extensively used to treat cancer. They attach themselves to antigens on cancer cells, or even non cancerous cells and other free-floating proteins. They can also act as immune checkpoint inhibitors [6]

    b. E.g., alemtuzumab. This is used in the treatment of multiple sclerosis and leukemia (CLL).

c. E.g., trastuzumab is a monoclonal antibody that acts against the HER2 protein and used in the treatment of carcinoma of the breast in which this protein is expressed in larger amounts on the surface of the cancer cells. It thereby causes the inactivation of the protein by blocking it [7].

2. **Conjugated monoclonal antibodies**

As the name suggests these are in combination with the chemotherapy drugs or radioactive materials. These are referred to as tagged or labeled mAbs. They directly deliver the therapy to the target cells causing minimal damage to the normal cells surrounding them after precisely identifying them [8]. It then delivers the toxic substance where it is needed most. They can be of the following types;

a. **Radiolabeled antibodies:** These are conjugated with radiolabeled particles. An excellent example is Ibritumomab tiuxetan which works against the CD20 antigen found on B lymphocyte cells. It is made up of radioactive substance (Yttrium-90). The mAb works on the target cancer cells and then the radioactive materials target the destined cells and also the nearby cell. *Radioimmunotherapy* (RIT) is the name used for this type of treatment [9].

b. **Chemolabeled antibodies:** These mAbs have chemotherapy drugs attached to them. Eg: Ado-trastuzumab emtansine, an antibody that targets the HER2 protein (breast cancer). It is covalently linked to the cytotoxic agent DM1

3. **Bispecific monoclonal antibodies**

These can attach to 2 different types of antigens at the same time, these have also been explored in cancer therapy and drug delivery. Example is blinatumomab, used in the treatment of acute lymphoblastic leukemia. It works by directing the body's T-cells (part of the immune system) to target and bind with the CD19 protein on the surface of B-cell lymphoma cells [10].

3. Monoclonal antibodies and prostate cancer

Prostate cancer is one of the most common cancers with its incidence being high in Americans but lesser in the Asian population. It develops within the prostate gland that is responsible for the production of seminal fluid. Cancer therapy considered for prostate cancer includes radical prostatectomy, radiation therapy, chemotherapy, brachytherapy and hormone therapy [11]. The role of mAbs in Prostate cancer treatment has not been very successful. Several trials have been carried out to check for its efficacy, the details of which have been mentioned below.

Ipilimumab was the first immune checkpoint inhibitor which received FDA approval for the treatment of metastatic melanoma. It worked as the anti-cytotoxic T lymphocyte-associated protein 4 (CTLA-4). This stimulated its exploration for the treatment of prostate cancer. Use of this mAb in conjugation with radiation therapy did show antitumor activity in the form of decreasing PSA levels. This was a phase 1 trial. Hence, phase 3 clinical trials were conducted for further evaluation where subjects were randomized to ipilimumab after chemotherapy or radiation therapy. These trials did show progression free survival but missed its endpoint of overall survival [12].

Clinical trials on another mAb Nivolumab remains under investigation. In a first of its kind combination immunotherapy using monoclonal antibodies, Ipilimumab plus nivolumab has been gaining responses as being reported in a phase 2 trial on metastatic castration resistant prostate cancer. Pembrolizumab is also an immune checkpoint inhibitor [13]. It has received approval from FDA for the treatment of prostate cancer. In these solid tumors, microsatellite instability (MSI) and mutations in mismatch repair genes (MMR) has been observed. Pembrolizumab is evaluated for in a patient after other effective treatments (such as sipuleucel-T, abiraterone, enzalutamide, docetaxel, cabazitaxel, radium-223, etc.) has been ruled out [14]. Combination therapies either with multiple immunotherapies or with immunotherapy and chemotherapy/RT, are currently being evaluated in prostate cancer. The optimal timing of immunotherapy in prostate cancer also remains unclear. Although much work remains to be done, the promise of prostate cancer immunotherapy remains unclear. There have been modern advances in the treatment of prostate cancer, however there is no curative treatment option once prostate becomes metastatic.

4. Monoclonal antibodies and renal cell carcinoma

Renal cell carcinoma is one of the urologic cancers that has lower incidence rates and poor prognosis. About 30% are diagnosed in their metastatic stage. It is a type of cancer that originates in the PCT. Therapy considered for this form of cancer include nephrectomy, radiation therapy, chemotherapy and embolization. The role of mAbs in Renal cell carcinoma is undertaken and studied in clinical trials treatment has not been very successful. Several trials have been carried out to check for its efficacy the details of which have been mentioned below as renal cell carcinoma (RCC) is a largely chemotherapy-resistant disease. It is immune responsive disease; therefore, checkpoint inhibitors can be considered as agents for the treatment of RCC [15].

The pivotal drug trial Checkmate 214 showed good objective responses in case of poor and intermediate risk patients in combination immunotherapy (Nivolumab/Ipilimumab) vs. the tyrosine kinase inhibitor sunitinib and can be considered as a first line treatment in these subjects for RCC. However, for favorable high-risk patients, the single agent sunitinib showed more response rate. Survival rates were similar in both arms. In another clinical trial Keynote 426, Pembrolizumab (anti-PD-1) plus axitinib, the responses were good and this led to its approval by leading to Food and Drug Administration (FDA) first line treatment. In another trial named, Javelin 101 Renal, avelumab (anti-PD-L1) plus axitinib vs. sunitinib the OS was not significantly different between the two arms [16].

In addition to this, there are many clinical trials that are underway for RCC (**Table 1**). Nivolumab was approved advanced clear-cell RCC by the FDA and is under investigation as pre- and postoperative therapy in mRCC. Combinatorial treatments with various drugs are also being studied in various clinical trials. Atezolizumab phase I trial involving 17 mRCC patients showed promising results as 7 had stable disease for more than 24 weeks. In another phase Ia study, of the 63 patients with clear-cell RCC whose OS was 28.9 months. Pembrolizumab is currently being investigated in two randomized phase II trials of mRCC patients. It has been found to be acceptable for safe use [17]. Several trials evaluating pembrolizumab in combination with various agents are also undergoing. Avelumab showed an acceptable usage when used in patients with advanced solid tumors and safety profile. Two ongoing trials are still being evaluated for avelumab in combination with axitinib Durvalumab. There are ongoing trials evaluating durvalumab in combination with other drugs, including tremelimumab for patients with advanced malignancies including RCC. Ipilimumab: Phase-II studies have been undergoing and the results are found to be partial response with adverse events being reported. In addition, Ipilimumab and nivolumab is being investigated and found to be favorable [18].

mAbs	Targeted therapy	Phase	Population
Nivolumab	Sunitinib Pazopanib	I	Advanced RCC, prior cytokine therapy allowed
Atezolizumab	Bevacizumab	Ib	Untreated, advanced clear cell RCC
Nivolumab	Bevacizumab	Neo-adjuvant pilot	Metastatic clear cell RCC, prior therapy allowed
Nivolumab	Temsirolimus	Ib/II	Metastatic RCC, prior therapy allowed
Pembrolizumab	Pazopanib	I/II	Untreated, advanced clear cell RCC
Pembrolizumab	Axitinib	Ib	Untreated, advanced clear cell RCC
Pembrolizumab	Bevacizumab	Ib/II	Metastatic clear cell RCC treated with failure of at least one prior therapy
Pembrolizumab	Aflibercept	I	Metastatic RCC treated with at least one prior VEGF TKI
Avelumab	Axitinib	Ib	Untreated, advanced clear cell RCC
Atezolizumab	Bevacizumab	III	Untreated, advanced clear cell RCC
Avelumab	Axitinib	III	Untreated, advanced clear cell RCC
Pembrolizumab	Axitinib	III	Untreated, advanced clear cell RCC

RCC: Renal cell carcinoma; VEGF: Vascular Endothelial Growth Factor; TKI: Tyrosine kinase inhibitors.

Table 1.
List of the Clinical trials that are underway for RCC.

5. Monoclonal antibodies and bladder cancer

Bladder cancer is one of the common cancers which develop in the tissues of the bladder. It is a type of Urolethial cancer. There are several methods which have been developed as a cancer therapy for bladder cancer and the most common being the, Bacillus Calmette-Guerin which has a very high success rate. The role of immunotherapy in Bladder cancer has been detailed in a number of case report and clinical trial studies [19]. The incidence of Bladder cancer is comparatively found to be higher in America when compared to other forms of malignancy.

Here are the various monoclonal antibodies that have been considered as a cancer therapy for the bladder cancer. 2016-Atezolizumab, was the first mAb to be approved by the FDA and also accepted by the European Association of Urology (EUA) as second-line therapy for patients with advanced Bladder Cancer. It is a PD-1/PD-L1 checkpoint inhibitor It has been used for subjects even with metastatic or advanced bladder cancer. 2017-Avelumab was also approved by FDA for urothelial cancers. It acts against PD-L1. A phase Ib clinical trial had been carried out with metastatic urothelial cancer which showed inconvincing results. However in the phase II trial, avelumab exhibited a good antitumor response in patients with advanced urothelial cancer whose tumors progressed during or after platinum-based chemotherapy 2017-Durvalumab has also received approval by FDA for the treatment of Bladder cancer. Studies in phases I and II patients have confirmed the effectiveness of durvalumab: It has shown responses in a number of clinical trial studies. It is a drug that acts against the PD-L1. 2017-Nivolumab is a FDA and EUA approved human mAb that acts against the PD-1. It was accepted on the basis of a single-arm phase trial for 270 platinum pretreated patients. The result has been 20% response rate [20].

Nivolumab was also tested on advanced or metastatic Bladder cancer subjects. In this study many adverse events were reported. Unlike the above mentioned mAbs,PD-L1 overexpression among patients was not significant. In another phase II clinical trial with subjects also receiving platinum-based chemotherapy showed a two-month progression-free period. In patients with PD-L1 overexpression compared to patients with low-expression, a difference in drug effects was observed. Many subjects did show adverse events [21].

Pembrolizumab has been showing positive responses in cases of advanced bladder cancer. It is a humanized monoclonal antibody used in the treatment of bladder cancer and is approved by the FDA and EAU. In a study conducted by on pembrolizumab by Bellmunt et al., it was observed that this mAb showed lower adverse events and longer survival by about 3 months which was significant when compared to chemotherapy drugs such as docetaxel and paclitaxel [22]. In a case report mentioning the treatment with pembrolizumab as reported by McDermott et al., it was observed that adverse events were not observed after 8 months and hence suggested that pembrolizumab can be considered as a PD –I inhibitor [23]. In patients with DNA repair defects, pembrolizumab can also be considered for treatment. This drug not only reduced the risk of developing newer cancers but also prevented premalignant hyperplastic lesion. This shall be a rational therapy. Pembrolizumab is also shown better survival rates when compared to chemotherapy as mentioned farina and his colleagues.

A novel murine monoclonal antibody KMP1 has been studied by cheng and his colleagues [24]. The study was conducted both in vitro and in vivo settings It identified the CD44 epitope on bladder cancer cells and bound to it due to O-linked glycosylation and thereby exhibit antitumor potential in both settings. Future studies may be recommended to understand the exact glycolsylation mechanism also produce humanized forms and also conjugate types for better therapeutics. Enfortumab vedotin delivers toxic drugs to tumors. It is an antibody-drug conjugate that targets the Nectin-4 pathway, it has been approved for further study in case of bladder cancer. Immunotherapy has significantly reduced the risk of recurrence for bladder cancer while also increasing the percentage of patients who see a complete response post-surgery. Investigational bladder cancer immunotherapies also have the capacity change the outcomes positively for patients with this disease.

6. Monoclonal antibodies and testicular cancer

Testicular cancer is a disease of the male organ, testicles that produces the male hormones and sperms. Approximately 90% of testicular cancer start in the germ cells which make the sperm and are referred to as the Germ cell tumor (GCT). They are of two types: seminomas and non-seminomas. The testicular cancer are the solid tumors that can be treated by chemotherapy even in the metastatic condition. However, the role of immunotherapy is still under investigation. The incidence of Testicular cancer has observed an increasing trend in both America and Europe [25].

6.1 Testicular cancer

There are several trials that have been directed towards the Testicular cancer. Several of these trials are against the Immune check point inhibitors. Many case studies have reported immune checkpoint inhibition efficacy in refractory GCTs. However, the mechanism by which this occurs is not clear. Trials have been conducted with mAbs, nivolumab or pembrolizumab (both anti-PD-1 agents) on subjects with refractory GCT. The results are very inconvincing on a phase II trial of Adra et *al*. [26] who administered pembrolizumab to 10 refractory

GCT patients. Despite of the PD-L1 status there were no responses and hence this led to the termination of the trial. Although three of seven patients with refractory germ cell tumors treated with nivolumab or pembrolizumab did show response, there was partial remission. Some case reports did mention about the rapid progression of the disease with pembrolizumab on single dose and some 40% response with single dose of nivolumab. In a case study reported by Chi and Schweizer, treatment response to nivolumab was observed hence, use of single checkpoint inhibitors have been unstable in nature [27]. No responses were observed by nadal et al. for a case report on a study conducted using Nibolumabwith cabozantini and bipilimumab [28]. Due to inefficacy of single agent durvalumab, the monotherapy arm was closed for a study conducted by Raggi et al. [29] Hence the results of immune checkpoint inhibitor monotherapy studies are disappointing and hence need more evaluation in many more clinical trials that shall be planned for future.

7. Monoclonal antibodies and penile cancer

Penile Cancer is a disease in which a tumor growth occurs in the tissues of the Penis. Although the localized penile cancer can be treated by penectomy, the metastatic forms need better strategies to deal with such as the standard Chemotherapy or the novel Immunotherapy. About 95% of the penile cancers are squamous cell carcinomas and other forms include the sarcoma, melanomas and the basal cell carcinoma. Most of the penile cancer is caused by HPV (human papilloma virus) infection. Although the incidence of Penile cancer is only about 1 in every 1,00,000 individuals I America and Europe, several Immunotherapy drug trials are underway to strategize its importance.

Epidermal growth factor receptor is usually overly expressed in Penile squamous cell carcinoma. EGFR amplification has been observed and thereby reported in a number of studies on primary penile squamous cell carcinoma. This amplification has been observed with poor prognosis in patients with penile squamous cell carcinoma and increased risk of recurrence. Considering this aspect, it has been chosen for treatment of systemic penile cancer. Immunotherapy towards EGFR target, includes monoclonal antibodies cetuximab and panitumumab [28]. In a study considering cetuximab either alone or with cisplatin, there was partial response. In another study considering cetuximab, panitumumab, and nimotuzumab about 50% of the patients showed response to treatment. However, this was a second line of treatment.

In addition to anti-EGFR therapy, immune checkpoint inhibitor drug trials of avelumab and pembrolizumab are under progress. These drugs are evaluating the role of PD- L1 and PD-1 inhibition with the above mentioned mAbs respectively, exclusively in penile carcinoma. The combination of PD-1 and cytotoxic T lymphocyte protein 4 (CTLA4) blockade might improve antitumour activity across multiple malignancies, including PSCC. However, the majority of the trials with patients suffering from penile carcinoma are basket trials that include because incidence of penile squamous cell carcinoma is very low [30]. In addition, Cetuximab, a chimeric monoclonal antibody is an epidermal growth factor receptor (EGFR) inhibitor and has still not received FDA approval for the treatment of penile cancer. Phase I trials of Nivolumab are also underway which is a conjunction of chemotherapy and lymphokine working against the HPV. As the frequency of this disease is very low, it has been difficult to conduct many trials. However, continual progress in the area of Immunotherapy with fewer trials has still been gaining approvals and success.

8. Conclusion

Immune status modification as strategy of cancer therapy does hold a significant place. Although, the conventional cancer treatments such as surgery, chemotherapy, and radiotherapy are still being referred to as the prominent ones, for some cancer types, immunotherapies are considered as first-line of treatment. One of the most important discoveries in the last several years in immunotherapy has been the development of immune checkpoint inhibitor, monoclonal antibodies that promote antitumor activity. T cells are a form of lymphocyte which are produced within the thymus and performs a crucial function in stimulating body's immune reaction to combat most cancers. They apprehend the overseas particles (antigens) with the aid of using particularly variable T cell receptor. Unlike antibody, the TCR cannot bind antigen and as a substitute it wishes to have peptides of the antigen proven to it with the aid of using an antigen presenting cell (APC). The molecules at the APC that gift the antigen are called as major histocompatibility complexes (MHC). Many stimulatory alerts also are wanted at this time. B7 is a form of peripheral membrane protein observed on activated antigen-providing cells (APC). This B7 on an APC can bind to cytotoxic T-lymphocyte-related antigen 4 (CTLA-4) developing an inhibitory sign and TCR activation. Once the activated T-cell receptor is within the tumor surroundings it is able to apprehend the antigen supplied with the aid of APC within the tumor. At this time, the programmed cell death protein 1 (PD-1) receptor also sends an inhibitory signal to the T-cell when the receptor binds to programmed cell death 1 ligand 1 (PD-L1), that's regularly expressed on tumor cells. Monoclonal antibodies act by inhibiting the binding PD-1 to PD-L1 and thereby boost body's immune response against the tumor cells.

Checkpoint inhibitors specifically goal PD-1/PD-L1) and CTLA4 Immune checkpoint efficacy is stricken by diverse factors, among which are tumor genomics, host germline genetics, PD-L1 levels, and intestine microbiome. Generally, in tumors, mutated or incorrectly expressed proteins are processed through the immunoproteasome into peptides which can be commonly loaded onto MHC molecules, which similarly now no longer usually are capable of eliciting CD8+ T cell reaction. This may also cause producing MHC-supplied immunogenic neoepitopes. It turned into proven, that after the intratumor heterogeneity rises, neoantigen-expressing clones emerge as greater homogenous with the differential expression of PD-L1.

There are number of FDA approved monoclonal antibodies, that are considered for the treatment of Urology oncology. These have been detailed in the **Table 1** and include FDA-approved PD-1 inhibitors such as nivolumab, pembrolizumab, cemiplimab, and FDA-approved PD-L1 such as atezolizumab, avelumab, and durvalumab. But all the open literatures do believe that combinatorial strategies with immune checkpoint therapies may provide a better survival benefit which have been demonstrated in various clinical trials. These can be in combination with radiation therapy, tyrosine kinase inhibitors and also many chemotherapeutic drugs. However, the response to immunotherapy with monoclonal antibodies varies subjectively and hence research into PD-L1 expression, gene signature expression, messenger RNA subtype, mutational and neoantigen load is essential to determine the varying response to monoclonal antibody immunotherapy. Although older modalities of treatment for cancer, has been extensively exploited, array of new drugs that offer hope of not only prolonging life but also curing significantly more patients in the future bring a ray of hope to the scientific world.

Conflict of interest

The authors declare conflict of interest as None.

Author details

Maya Kulshekar[1], Shridhar C. Ghagane[2]*, Sridevi I. Puranik[3], Rajendra B. Nerli[4] and Murigendra B. Hiremath[5]

1 Department of Biochemistry, JN Medical College, KLE Academy of Higher Education and Research (Deemed-to-be-University), JNMC Campus, Belagavi, Karnataka, India

2 Division of Urologic-Oncology, Department of Urology, KLES Dr. Prabhakar Kore Hospital and Medical Research Centre, Belagavi, Karnataka, India

3 Department of Zoology, KLES B.K. Arts, Science and Commerce College, Chikodi, Karnataka, India

4 Department of Urology, JN Medical College, KLE Academy of Higher Education and Research (Deemed-to-be-University), JNMC Campus, Belagavi, Karnataka, India

5 Department of Biotechnology and Microbiology, Karnatak University, Dharwad, Karnataka, India

*Address all correspondence to: shridhar.kleskf@gmail.com

References

[1] Bander NH, Nanus DM, Milowsky MI, Kostakoglu L, Vallabahajosula S, Goldsmith SJ. Targeted systemic therapy of prostate cancer with a monoclonal antibody to prostate-specific membrane antigen. InSeminars in oncology 2003 Oct 1 (Vol. 30, No. 5, pp. 667-676). WB Saunders.

[2] Vallabhajosula S, Goldsmith SJ, Kostakoglu L, Milowsky MI, Nanus DM, Bander NH. Radioimmunotherapy of prostate cancer using 90Y-and 177Lu-labeled J591 monoclonal antibodies: effect of multiple treatments on myelotoxicity. Clinical cancer research. 2005;11(19):7195s–200s.

[3] Gong MC, Chang SS, Sadelain M, Bander NH, Heston WD. Prostate-specific membrane antigen (PSMA)-specific monoclonal antibodies in the treatment of prostate and other cancers. Cancer and Metastasis Reviews. 1999;18(4):483-490.

[4] Azevedo R, Ferreira JA, Peixoto A, Neves M, Sousa N, Lima A, Santos LL. Emerging antibody-based therapeutic strategies for bladder cancer: A systematic review. Journal of Controlled Release. 2015; 214:40-61.

[5] Sanz L, Cuesta ÁM, Compte M, ÁLVAREZ-VALLINA L. Antibody engineering: facing new challenges in cancer therapy 1. Acta Pharmacologica Sinica. 2005; 26(6):641-648.

[6] Kiessling A, Füssel S, Wehner R, Bachmann M, Wirth MP, Rieber EP, Schmitz M. Advances in specific immunotherapy for prostate cancer. European urology. 2008; 53(4):694-708.

[7] Ross JS, Gray KE, Webb IJ, Gray GS, Rolfe M, Schenkein DP, Nanus DM, Millowsky MI, Bander NH. Antibody-based therapeutics: focus on prostate cancer. Cancer and Metastasis Reviews. 2005; 24(4):521-537.

[8] Silvestri I, Cattarino S, Agliano AM, Collalti G, Sciarra A. Beyond the immune suppression: the immunotherapy in prostate cancer. BioMed research international. 2015; 2015.

[9] Christiansen J, Rajasekaran AK. Biological impediments to monoclonal antibody–based cancer immunotherapy. Molecular cancer therapeutics. 2004; 3(11):1493-1501.

[10] Lange PH, Winfield HN. Biological markers in urologic cancer. Cancer. 1987; 60(S3):464-472.

[11] Takeda K, Okumura K, Smyth MJ. Combination antibody-based cancer immunotherapy. Cancer science. 2007; 98(9):1297-1302.

[12] Mehta K, Patel K, Parikh RA. Immunotherapy in genitourinary malignancies. Journal of hematology & oncology. 2017;10(1):1-4.

[13] Tsiatas M, Grivas P. Immunobiology and immunotherapy in genitourinary malignancies. Annals of translational medicine. 2016; 4(14).

[14] Baxevanis CN, Perez SA, Papamichail M. Combinatorial treatments including vaccines, chemotherapy and monoclonal antibodies for cancer therapy. Cancer immunology, immunotherapy. 2009; 58(3):317-324.

[15] Hekman MC, Rijpkema M, Muselaers CH, Oosterwijk E, Hulsbergen-Van de Kaa CA, Boerman OC, Oyen WJ, Langenhuijsen JF, Mulders PF. Tumor-targeted dual-modality imaging to improve intraoperative visualization of clear cell renal cell carcinoma: a first in man study. Theranostics. 2018;8(8):2161.

[16] Choueiri TK, Larkin J, Oya M, Thistlethwaite F,

Martignoni M, Nathan P, Powles T, McDermott D, Robbins PB, Chism DD, Cho D. Preliminary results for avelumab plus axitinib as first-line therapy in patients with advanced clear-cell renal-cell carcinoma (JAVELIN Renal 100): an open-label, dose-finding and dose-expansion, phase 1b trial. The lancet oncology. 2018;19(4):451-460.

[17] Xu JX, Maher VE, Zhang L, Tang S, Sridhara R, Ibrahim A, Kim G, Pazdur R. FDA approval summary: nivolumab in advanced renal cell carcinoma after anti-angiogenic therapy and exploratory predictive biomarker analysis. The oncologist. 2017; 22(3):311.

[18] Atkins MB, Jegede O, Haas NB, McDermott DF, Bilen MA, Drake CG, Sosman JA, Alter RS, Plimack ER, Rini BI, Hurwitz ME. Phase II study of nivolumab and salvage nivolumab+ ipilimumab in treatment-naïve patients (pts) with advanced renal cell carcinoma (RCC) (HCRN GU16-260). 2020: 5006-5006.

[19] Fuge O, Vasdev N, Allchorne P, Green JS. Immunotherapy for bladder cancer. Research and reports in urology. 2015; 7:65.

[20] Bellmunt J, Powles T, Vogelzang NJ. A review on the evolution of PD-1/ PD-L1 immunotherapy for bladder cancer: the future is now. Cancer treatment reviews. 2017; 54:58-67.

[21] Massari F, Di Nunno V, Cubelli M, Santoni M, Fiorentino M, Montironi R, Cheng L, Lopez-Beltran A, Battelli N, Ardizzoni A. Immune checkpoint inhibitors for metastatic bladder cancer. Cancer treatment reviews. 2018; 64:11-20.

[22] Bellmunt J, De Wit R, Vaughn DJ, Fradet Y, Lee JL, Fong L, Vogelzang NJ, Climent MA, Petrylak DP, Choueiri TK, Necchi A. Pembrolizumab as second-line therapy for advanced urothelial carcinoma. New England Journal of Medicine. 2017; 376(11):1015-1026.

[23] McDermott J, Jimeno A. Pembrolizumab: PD-1 inhibition as a therapeutic strategy in cancer. Drugs of today (Barcelona, Spain: 1998). 2015; 51(1):7-20.

[24] Chen Y, Wang H, Zuo Y, Li N, Ding M, Li C. A novel monoclonal antibody KMP 1 has potential antitumor activity of bladder cancer by blocking CD 44 in vivo and in vitro. Cancer medicine. 2018; 7(5):2064-2077.

[25] Masiuk M, Lewandowska M, Teresinski L, Dobak E, Urasinska E. Nucleolin and nucleophosmin expression in seminomas and non-seminomatous testicular tumors. Folia histochemica et cytobiologica. 2019;57(3):139-145.

[26] Adra N, Althouse SK, Ammakkanavar NR, Radovich M, Albany C, Vaughn DJ, Einhorn LH, Hanna NH. Phase II trial of pembrolizumab in patients (pts) with incurable platinum refractory germ cell tumors (GCT). 2017: 4520-4520.

[27] Chi EA, Schweizer MT. Durable Response to Immune Checkpoint Blockade in a Platinum-Refractory Patient with Nonseminomatous Germ Cell Tumor. Clinical genitourinary cancer. 2017; 15(5): e855–e857.

[28] Nadal RM, Mortazavi A, Stein M, Pal SK, Davarpanah NN, Parnes HL, Ning YM, Cordes LM, Bagheri MH, Lindenberg L, Thompson R. Results of phase I plus expansion cohorts of cabozantinib (Cabo) plus nivolumab (Nivo) and CaboNivo plus ipilimumab (Ipi) in patients (pts) with metastatic urothelial carcinoma (mUC) and other genitourinary (GU) malignancies. 2018: 515-515.

[29] Raggi D, Necchi A, Giannatempo P. Nivolumab and its use in the second-line treatment of metastatic urothelial

cancer. Future Oncology. 2018; 14(26):2683-2690.

[30] Trafalis DT, Alifieris CE, Kalantzis A, Verigos KE, Vergadis C, Sauvage S. Evidence for efficacy of treatment with the anti-PD-1 Mab nivolumab in radiation and multichemorefractory advanced penile squamous cell carcinoma. Journal of Immunotherapy. 2018; 41(6):300-305.

Chapter 7

Emerging Monoclonal Antibodies for the Treatment of Multiple Myeloma

Hanley N. Abramson

Abstract

Therapeutic measures designed to treat multiple myeloma (MM) have undergone a fundamental shift over the past two decades as a number of small molecules that attack this cancer by different mechanisms, including proteasome blockade, immunomodulation, and histone deacetylase (HDAC) inhibition, have been introduced. The insertion of monoclonal antibodies (mAbs) into the mix began in 2015 with the U.S. Food and Drug Administration (FDA) approval of daratumumab and elotuzumab, which target CD38 and SLAMF7, respectively. In 2020, they were joined by another anti-CD38 mAb, isatuximab, and the bispecific antibody-drug conjugate (ADC) belantamab mafodotin, which targets the B-cell maturation antigen (BCMA). This review focuses on additional mAbs currently under clinical study for MM. These include several BCMAxCD3-directed bispecifics (AMG 420, AMG 701, REGN5458, REGN5459, teclistamab, and TNB-383B), the ADCs indatuximab ravtansine and STRO-001, and checkpoint inhibitors, although the future status of the latter is in a state of flux due to toxicity issues that arose in trials in which these drugs, especially PD-1 or PD-L1 blockers, were combined with immunomodulators.

Keywords: CD38, B-cell maturation antigen, daratumumab, elotuzumab, isatuximab, belantamab mafodotin, indatuximab ravtansine, STRO-001

1. Introduction

Multiple myeloma (MM), a malignancy of plasma cells, ranks second among all blood cancers in the U.S., representing about 1% of all diagnosed malignancies. In 2020, an estimated 32,270 individuals (54.3% male), the majority over age 65, will be diagnosed with the disease and approximately 12,830 will succumb to it [1]. Classical symptoms of active MM include hypercalcemia, renal insufficiency, anemia, and bone lesions (CRAB) and often are preceded by an asymptomatic stage referred to as monoclonal gammopathy of undetermined significance (MGUS). The risk of progression from MGUS to MM is about 1% per year [2] and may include another asymptomatic state known as smoldering myeloma [3]. Current guidelines for the diagnosis and treatment of MM have been published by the National Comprehensive Cancer Network (NCCN) [4]. Treatment modalities for MM have seen enormous advances since the beginning of the current century with the introduction of drugs working by different mechanisms, e.g., proteasome inhibition (bortezomib, carfilzomib, and ixazomib) and immunomodulation (lenalidomide and pomalidomide), which were added to the long-established treatments based on alkylating agents (melphalan and cyclophosphamide) and corticosteroids (dexamethasone). These measures, together with autologous stem cell transplantation (ASCT), first introduced for MM in the 1990's, have increased the five-year survival rate for the disease from 24% in the mid-1970s to 55% in the 2010–2016 period [5]. Furthermore, the relatively recent arrival on the scene of monoclonal antibodies (mAbs), beginning with the U.S. Food and Drug Administration (FDA) approval of daratumumab in 2015, has greatly expanded the therapeutic options available to treat MM. However, in spite of these advances, MM remains incurable as patient relapse and refractoriness to treatment continue as major issues. This review focuses on the contributions made by those mAbs currently approved for MM, as well as on those under investigation as potential future therapies for this disease.

2. mAbs targeting CD38

CD38 is a multifunctional 45 kDa type II transmembrane glycoprotein, lacking an internal signaling domain, that is expressed at high levels on both malignant and normal plasma cells and has attracted much interest as a target for drug development in MM [6]. It also is found normally, but at lower levels, on the surfaces of T and B lymphocytes, natural killer (NK) cells, and monocytes. Among its several roles, CD38 acts as a receptor for CD31 (platelet endothelial cell adhesion molecule; PECAM-1) [7] and as a cyclic ADP ribose hydrolase, an ectoenzyme whose reaction products play an essential role in regulation of intracellular calcium levels [8].

Antibodies directed against CD38 kill myeloma cells by a number of possible mechanisms, chief among them being antibody-dependent cellular cytotoxicity (ADCC), antibody-dependent cellular phagocytosis (ADCP), and complement-dependent cytotoxicity (CDC). Intracellular signal cascade disruption, the result of crosslinks formed between myeloma cell CD38 and Fcγ receptors on effector cells, may also play an important role in initiating apoptotic events in myeloma cells [9]. In addition, anti-CD38 antibodies have been shown to exhibit immunomodulatory effects that cause blockage of regulatory T- and B-cells and myeloid-derived suppressor cells [10].

Daratumumab (Darzalex®), a fully human IgG1κ mAb targeting CD38, initially was approved for the management of MM in patients who had relapsed following at least three prior therapies including an immunomodulator and a proteasome inhibitor [11]. Approval was supported by the results of two phase III trials - POLLUX (NCT02076009) and CASTOR (NCT02136134) - in which daratumumab/dexamethasone was combined respectively with either lenalidomide [12] or bortezomib [13]. Further encouraging data from phase III trials, demonstrating deeper and more sustained responses combined with good tolerability, soon enabled daratumumab/corticosteroid combinations with immunomodulators or proteasome inhibitors to assume an important role in even earlier courses of treatment [14, 15], as well as in newly diagnosed patients, whether ASCT-eligible [16] or -ineligible [17, 18]. Several network meta-analytic studies of random controlled trials covering a number of different settings, including in patients with newly diagnosed disease, have demonstrated the benefits of daratumumab-containing regimens in MM therapy with respect to efficacy and safety [19]. Furthermore, several reports have indicated the efficacy of daratumumab monotherapy in patients who have failed earlier lines of anti-myeloma therapy [20], as well as in patients with smoldering MM [21]. Also, the FDA recently has approved a subcutaneous formulation of daratumumab plus hyaluronidase, which enables shorter infusion times without compromising safety or efficacy [22].

Isatuximab (Sarclisa®, SAR650984) is a chimeric mouse-human IgG1κ CD38-targeting mAb that was approved by the FDA in March 2020 for the treatment of relapsing and/or refractory MM (RRMM) in combination with pomalidomide and dexamethasone in patients who have received at least two prior therapies, including lenalidomide and a proteasome inhibitor [23]. Approval was predicated primarily on the results of the phase III ICARIA-MM trial (NCT02990338) in which addition of isatuximab to a pomalidomide/dexamethasone regimen resulted in a five month increase in median progression free survival (PFS), from 6.5 to 11.5 months [24]. Upper respiratory infections and diarrhea were the most frequently encountered adverse events noted in both groups. Although infusion reactions (mostly, grades 1 and 2) were reported in 38% of patients in the isatuximab cohort, this reaction was not noted in any patients in the pomalidomide/dexamethasone group. An additional five phase III trials that include isatuximab/dexamethasone currently are in progress for: newly diagnosed MM (NDMM) (NCT03617731 and NCT03319667 IMROZ - lenalidomide/bortezomib; NCT04483739 - lenalidomide/carfilzomib); RRMM (NCT03275285 IKEMA - carfilzomib); and high-risk smoldering MM (NCT04270409 - lenalidomide).

The mechanism of action of isatuximab exhibits some significant differences from that of daratumumab. For example, the former appears to work principally through ADCC with only minor contributions from CDC [25]. Also, unlike daratumumab, crosslinking induced by isatuximab is not a prerequisite for initiation of target cell apoptosis [26]. Moreover, isatuximab is a much more potent inhibitor of ectoenzyme activity although the significance of this is unknown [27].

MOR202 (MOR03087, TJ202) is a fully humanized IgG1λ mAb that exhibited an objective response rate (ORR) of 29% in a phase II trial with dexamethasone in patients who had previously received four lines of therapy [28]. In addition, this drug has shown some promising efficacy when combined with immunomodulators [29, 30]. MOR202 also appears to offer the advantage of requiring reduced infusion times and is associated with reduced infusion-related reactions compared to daratumumab or isatuximab, possibly due to its lack of dependency on CDC as a function of its activity. However, the drug's sponsor, MorphoSys AG, recently decided to discontinue further development of MOR202 for MM. Two additional anti-CD38 mAbs from Takeda Oncology currently are in the early stages of clinical development for RRMM: TAK-573, an IgG4 antibody conjugated to an attenuated form of interferon α [31], and TAK-079, a fully humanized IgG1λ mAb [32].

3. mAbs targeting SLAMF7

A group of surface proteins belonging to the signaling lymphocytic activation molecule family (SLAMF) has elicited considerable interest in recent years due to the high expression of four family members (SLAMF2, 3, 6, and 7) on both normal plasma cells and those from MM patients at all stages of disease. No trials of SLAMF2-targeting mAbs have been initiated and clinical studies of the SLAMF3 and SLAMF6 mAbs SGN-CD48A and azintuxizumab vedotin (ABBV-838), respectively, both were halted early in phase I trials. On the other hand, SLAMF7 (CS1 or CD319) has emerged as the principal focus for new anti-myeloma mAb development in this group of targets with the introduction of elotuzumab (Empliciti®), a humanized IgG1κ mAb [33]. Preclinical studies revealed that the anti-myeloma activity of elotuzumab is the result of ADCC involving direct activation and engagement of NK cells [34]. FDA approval in 2015 of elotuzumab, which lacks activity as a single agent, was the result of the ELOQUENT-2 trial (NCT01239797) involving 646 randomly assigned RRMM patients who received the mAb plus dexamethasone with or without lenalidomide. The cohort receiving elotuzumab exhibited a PFS of 19.4 months and an ORR of 68% at one year and 41% at two years, compared to 14.9 months and 57% and 27% for the control [35]. These results were confirmed further by a subsequent four-year follow-up study [36]. Similar benefits of elotuzumab in RRMM were observed in combination with pomalidomide-dexamethasone in the ELOQUENT-3 trial (NCT02654132) [37], which included patients refractory to both lenalidomide and a proteasome inhibitor and resulted in the 2018 FDA approval of this combination [38]. Favorable data also have been generated in a trial (NCT01478048) in which bortezomib-dexamethasone was included with elotuzumab [39].

4. BCMA-targeting antibody-drug complexes

The cytokines BAFF (B-cell activating factor) and APRIL (a proliferation-inducing ligand) have received much attention in recent years for their roles in the pathophysiology of autoimmune diseases [40]. In addition, there is evidence that these two homologous members of the tumor necrosis factor (TNF) superfamily play roles in myeloma cell viability and proliferation [41]. Two other TNF family members - transmembrane activator and calcium modulator and cyclophilin ligand interactor (TACI) and B-cell maturation antigen (BCMA), located on the surface of myeloma cells, serve as ligands for both BAFF and APRIL [42]. While the respective BAFF and APRIL inhibitors, atacicept and tabalumab (LY2127399), as well as the anti-APRIL mAb BION-1301, showed poor efficacy in MM trials [43–45], BCMA has surfaced as a compelling target in anti-myeloma drug research [46].

BCMA normally plays a key role in B-cell differentiation into plasma cells [47]. Myeloma cells, whether from cell lines or patient samples, exhibit not only consistent and virtually exclusive elevation of BCMA levels and its mRNA during malignant transformation but also at similar levels during the various stages of MM from previously untreated to relapse [48]. A soluble form of BCMA (sBCMA), which results from the shedding of BCMA from the plasma cell surface by the action of γ-secretase, is an important factor that, by lowering the density of the target antigen while also providing a soluble decoy, potentially limits the efficacy of BCMA blockers in clinical development, accounting for the inclusion of γ-secretase inhibitors in a number of BCMA-targeted trials [49].

Removal of several of the fucosyl groups normally found in the N-linked biantennary complex oligosaccharides in the Fc region of IgG antibodies is a well-established approach for enhancing ADCC through binding of FcγIIIa receptors on NK cells [50]. One such anti-BCMA mAb is the afucosylated antibody-drug conjugate (ADC) belantamab mafodotin (Blenrep®, GSK2857916), in which the antibody is coupled to the microtubule inhibitor monomethylauristatin F (MMAF) through a protease-resistant maleimidocaproyl linker. While the antibody component disrupts BAFF/APRIL myeloma cell signaling by binding to the BCMA receptor to induce ADCC, the MMAF component causes cell cycle arrest at the G2/M interface [51]. Belantamab mafodotin continues to be the subject of the DREAMM series of trials in RRMM patients. An early exploratory study (NCT02064387, DREAMM-1) found an ORR of 60% in 35 heavily pre-treated RRMM patients when the immunoconjugate was used as a single agent [52]. This encouraging response level dropped to 31% (30/97) in RRMM patients refractory to proteasome inhibitors, immunomodulators, and/or anti-CD38 therapy, who received the drug as monotherapy at 2.5 mg./Kg. in the ensuing phase II DREAMM-2 trial (NCT03525678) [53]. However, the efficacy level was considered comparable to that observed with other therapies for RRMM patients with similar numbers of prior therapies. Analysis of adverse event data in the DREAMM-2 trial concluded that belantamab mafodotin exhibits an acceptable safety profile with ocular toxicity, primarily in the form of keratopathy as the most commonly reported adverse event and attributable to the microtubule-inhibitor payload [54], presenting the greatest concern. Based on these data, in August, 2020, belantamab mafodotin was approved by the FDA, under the provisions of accelerated review, as monotherapy for RRMM patients who have received at least four prior treatments that included an immunomodulator, a proteasome inhibitor, and an anti-CD38 monoclonal antibody [55]. In addition to these two milestone studies, a phase II trial (NCT04126200; DREAMM-5) has been initiated that includes belantamab mafodotin monotherapy with two T-cell costimulatory agonist mAbs – the OX40-targeting GSK3174998 and the inducible co-stimulator (ICOS) GSK3359609, along with nirogacestat (PF-03084014), a γ-secretase inhibitor [56]. The ADC plus low-dose dexamethasone also is incorporated into four phase III investigations that include standard therapies such as: pomalidomide (NCT04162210; DREAMM-3, RRMM); bortezomib and daratumumab (NCT04246047; DREAMM-7, RRMM); pomalidomide and bortezomib (NCT04484623; DREAMM-8, RRMM); and lenalidomide and bortezomib (NCT04091126; DREAMM-9, NDMM) [57].

Another BCMA-targeted ADC that has elicited much interest for its anti-myeloma action is MEDI2228, which is comprised of a fully human mAb attached to a minor-groove binding pyrrolobenzodiazepine (tesirine) payload via a protease-cleavable valine-alanine linker [58]. Release of the warhead following internalization of the immunoconjugate and trafficking to the lysosome results in DNA damage and subsequent apoptosis. Preclinical studies in mice revealed this agent's potent anti-myeloma activity even when clinically significant levels of sBCMA were present [59]. Currently, MEDI2228 is the subject of a phase I clinical trial (NCT03489525) to determine appropriate dosing as monotherapy in RRMM patients; however, no results have been reported as yet. Another anti-BCMA ADC that has entered clinical studies for MM is AMG 224 (NCT02561962), comprised of a maytansine analog connected to a non-cleavable 4-(*N*-maleimidomethyl) cyclohexane-1-carboxylate linker [60]. CC-99712 is yet another BCMA-targeted ADC (undisclosed composition) that recently entered a clinical trial (NCT04036461) for RRMM. Other anti-BCMA ADC mAbs that have demonstrated promise in preclinical work but for which human studies have not yet begun include two proprietary products, referred to as BCMA-077 and BCMA-024 [61], and HDP-101, in which the conjugate is the potent RNA polymerase II subunit A (POLR2A) inhibitor α-amanitin [62].

5. T-cell-engaging bispecific antibodies

The T-cell-engaging bispecific antibody (T-BsAb) concept, originally developed by Nisonoff in 1961 [63], is based on the design of a dual-targeting antibody whereby one arm initially binds to the T-cell CD3 co-receptor complex while the other arm is subsequently directed to a tumor-associated antigen. The immunological synapse created between the two cells causes release of two cytolytic-initiating proteins: perforin, which causes formation of transmembrane pores in the malignant cell and granzyme B, which traverses the pores thus produced to initiate tumor cell apoptosis. The T-BsAb strategy differs from normal T-cell mediated cytotoxicity by removing requirements for costimulatory signals, formation of an antigen-major histocompatibility complex (MHC), and for *ex vivo* T-cell manipulation, thus permitting the possibility of "off-the-shelf" product manufacture. Furthermore, persistent T-cell activation enables polyclonal expansion of T memory cells. In addition, the therapeutically relevant properties of constructs may be fine-tuned to optimization by altering biodistribution characteristics and the relative binding affinities of each arm for their respective targets [64].

Amgen's proprietary BiTE® platform represents an innovative subclass of T-BsAb in which tandem single-chain variable fragments (scFvs) provide the cross-link [65]. The first successful application of BiTE® technology was the CD3-CD19 cross-linking construct blinatumomab (Blincyto®), which was approved by the FDA in 2014 for Philadelphia chromosome-negative B-cell precursor acute lymphocytic leukemia (B-cell ALL). Although a single trial (NCT03173430) of blinatumomab in RRMM patients had to be terminated because of "slow patient accrual", the majority of myeloma-related work using BiTE® constructs have been based on recombinant antibodies designed to cross-link surface tumor-specific T-cell CD3ζ chains and targeted myeloma cell BCMA.

Initial results with the BCMAxCD3 BiTE® product AMG 420 (BI-836909), which was accorded fast-track status by the FDA in 2018, showed an ORR of 31% in 42 RRMM patients, including seven of ten patients refractory to at least two lines of therapy who received the maximum tolerated dose of 400 μg/day. Infections and polyneuropathy were the most serious adverse events noted in this trial. Cytokine release syndrome (CRS; cytokine storm), predominantly grade 1, was observed in 38% of patients in the study [66, 67]. Monotherapy with AMG 701, a related BiTE® construct with a longer serum half-life than AMG 420, currently is the focus of a phase I trial (NCT03287908) for RRMM. Data generated in a preclinical investigation suggests that future consideration of a trial of AMG 701 in combination with an immunomodulator may be warranted [68]. In addition, two other BCMAxCD3 bispecific antibodies from Regeneron, REGN5458 (NCT03761108) and REGN5459 (NCT04083534), are in phase I RRMM studies. Preliminary data on the first three patients treated with the former agent have been reported [69].

Following favorable safety results in a monkey model [70], two phase I RRMM clinical trials of the BCMAxCD3 bispecific teclistamab (JNJ-64007957) have been initiated. In addition to a dose-escalation study (NCT03145181), this agent has been incorporated into a trial (NCT04108195) that includes subcutaneous daratumumab and talquetamab (JNJ-64407564), a CD3xGPRC5D bispecific construct. Another BCMAxCD3 formatted product under scrutiny for RRMM is PF-06863135 (PF-3135) (NCT03269136), the result of hinge-mutation engineering of an IgG2a backbone [71]. CC-93269, a T-cell engager whose arms bind in a 2 + 1 arrangement, monovalently to CD3ε and bivalently to BCMA, is another member of this class in a myeloma-based trial (NCT03486067) [72].

TNB-383B, a BCMAxCD3 T-BsAb resulting from collaboration between Tenebio and Abbvie and designated as an orphan drug by the FDA, differs from other drugs in this class in that its structure consists of a single immunoglobulin light chain and two variable heavy chains. The product, which recently began a phase I trial for RRMM (NCT03933735), is noteworthy for its strong T-cell activation kinetics and low affinity anti-CD3 arm, which results in reduced release of cytokines while retaining high cytotoxicity toward myeloma cells [73]. A number of other bispecific antibodies have exhibited promise for RRMM in preclinical work. These include TNB-381 M [74], FPA-151 [74], EM801 [48], and AP163 [75]. HPN217, developed by Harpoon Therapeutics, is a tri-specific antibody possessing three binding domains in a single chain – a C-terminal single-chain CD3ε T-cell receptor-(TCR)-binding component, a human serum albumin-binding central domain, and an N-terminal BCMA-binding portion. This product, which is in a phase I/II trial for RRMM (NCT04184050), has an extended half-life compared to bispecific formats, a property ascribed to its smaller size and flexibility [76]. Moreover, bispecifics based on myeloma surface antigens other than BCMA have been developed as alternative CD3 epitope binding partners. In addition to the aforementioned talquetamab (NCT03399799) [77], these include the CD3xCD38 construct found in GBR 1342 (NCT03309111) [78] and the CD3xFcRH5 design incorporated into BFCR4350A (NCT03275103) [79].

In addition to the BCMAxCD3 bispecific formats noted above, creation of BCMA-targeted constructs directed to receptors on NK cells has been described. Like cytotoxic T-cells, NK cells are known to mediate cytotoxicity through a variety of mechanisms, including granzyme-perforin release and through expression of various apoptosis-inducing ligands [80]. A tri-specific product that binds both BCMA and CD200 on myeloma cells to CD16A on NK cells represents one such drug [81]. Similarly, CTX-4419, which binds BCMA to both NK cell CD16A and p30, has shown initial promise in preclinical models although CD16A binding does not appear to be a requirement for the anti-myeloma activity of this product [82]. Other BCMA-NK cell-engaging antibodies, such as CTX-8573 [83] and AFM26 [84], also have shown some potential as RRMM therapies.

6. Immune checkpoint inhibitors

Over the past decade, immune checkpoint blockade has emerged as a principal strategy for new antitumor drug development. This immunotherapeutic approach is based on identification of biomarkers and their cognate ligands that enable the body's immune system to overcome the capacity of tumor cells to evade immune surveillance and elimination, as well as on the design of mAbs to block these interactions. In its most successful application to date, discoveries made concerning the roles played in this process by cytotoxic T-lymphocyte–associated protein-4 (CTLA-4), the first member of this class to serve as a clinical target, and the programmed death (PD) receptor, have given momentum to this innovative line of attack on a variety of tumor types [85].

Costimulatory signals resulting from interaction of CD28 on the surface of T-cells with its CD80 (B7–1) ligand on antigen-presenting cells play an important role in activating T-cells. CTLA-4, expressed on the T-cell plasma membrane, competes with CD80-CD28 binding to downregulate T-cell activation and thus represents an important mechanism that suppresses immune responses and, as a consequence, enables tumor cells to evade detection. CTLA-4-directed mAbs, by competing with the CD80-CD28 interaction, enhance the ability of T-cells to generate an antitumor response. This strategic approach was successfully applied to the immunotherapy of advanced melanoma by the anti-CTLA-4 mAb ipilimumab, which was approved by the FDA in 2011 and has been extended since to include a number of other solid tumors [86]. However, the drug has shown less than impressive results in hematologic cancers, such as acute myeloid leukemia [87]. One trial (NCT02681302) of ipilimumab combined with nivolumab that included both lymphoma and MM patients is currently active but so far only preliminary efficacy and toxicity data have been reported [88].

Binding of PD-1, expressed on the surface of T-cells, to PD-L1 or PD-L2 on tumor cells inhibits cytotoxic T lymphocyte proliferation and cytokine secretion while also causing an increase in the T regulatory cell population. These combined effects produce immune tolerance, enabling unrestrained tumor cell growth and survival [89]. Since 2014, the FDA has approved three PD-1 inhibitors – pembrolizumab (Keytruda®), nivolumab (Opdivo®), and cemiplimab (Libtayo®) and three PD-L1 blockers – avelumab (Bavencio®), durvalumab (Imfinzi®), and atezolizumab (Tecentriq®). Pembrolizumab, the first to be approved and the most versatile member of the checkpoint blocker group, has been approved for

17 different indications, many as front-line therapy for solid tumors ranging from melanoma to small-cell lung cancer to metastatic Merkel cell carcinoma [90]. In 2017, the FDA in an unprecedented move approved pembrolizumab for the treatment of solid tumors having a microsatellite instability (mismatched repair

Trial ID [reference]	N	Trial title
PD-1 Inhibitors		
NCT03848845 [102]	40	A phase I/II single arm open-label study to explore safety and clinical activity of GSK2857916 administered in combination with pembrolizumab in subjects with relapsed/refractory multiple myeloma (DREAMM 4)
NCT03506360	41	Phase II trial of pembrolizumab, ixazomib, and dexamethasone for relapsed multiple myeloma
NCT04361851	33	Phase II study of daratumumab-pembrolizumab for multiple myeloma patients with ≥ three prior lines of therapy
NCT03168438 [103]	20	NY-ESO-1^{c259}T alone and in combination with pembrolizumab for multiple myeloma
NCT03782064	25	A phase II trial of vaccination with dendritic cell (DC)/myeloma fusions in combination with nivolumab in patients with relapsed multiple myeloma
NCT03292263 [104]	30	Autologous stem cell transplantation with nivolumab in patients with multiple myeloma
NCT04119336	50	A phase II study of nivolumab in combination with ixazomib, cyclophosphamide, and dexamethasone in relapsed and refractory multiple myeloma
NCT03194867	109	A phase I/II study to evaluate safety, pharmacokinetics and efficacy of isatuximab in combination with cemiplimab in patients with relapsed/refractory multiple myeloma
NCT03111992	26	Phase I/Ib, multi-center, open-label, study of single agent CJM112 (anti-IL17A mAb), and spartalizumab (PDR001) in combination with LCL161 (SMAC mimetic) or CJM112 in patients with relapsed and/or refractory multiple myeloma
PD-L1 Inhibitors		
NCT02431208 [105, 106] Trial Start: 7/22/15	300	A phase Ib study of the safety and pharmacokinetics of atezolizumab alone or in combination with an immunomodulatory drug and/or daratumumab in patients with multiple myeloma (relapsed/ refractory and post-autologous stem cell transplantation)
NCT03910439	30	A phase II pilot study of avelumab in combination with hypofractionated radiotherapy in patients with relapsed refractory multiple myeloma
CD47 Inhibitors		
NCT03530683	156	A phase Ia/Ib dose escalation and expansion trial of TTI-622 in patients with advanced relapsed or refractory lymphoma or myeloma
LAG-3 and TIGIT Inhibitors		
NCT04150965	104	A phase I/II assessment of combination immuno-oncology drugs elotuzumab, anti-LAG-3 (BMS-986016) and anti-TIGIT (BMS-986207)

Trial start date on or after July 1, 2017, unless otherwise noted.

Table 1.
Selected active trials of checkpoint inhibitors in MM.

deficiency). Known as a tissue agnostic approach to cancer diagnosis and treatment, this marked the first drug approval based on a specific biomarker instead of on the tissue or organ of origin [91].

In terms of hematologic malignancies, pembrolizumab has been approved for treatment of Hodgkin's lymphoma while both pembrolizumab and nivolumab have been approved for primary mediastinal large B-cell lymphoma (PMBCL). While studies of checkpoint inhibitors in both chronic and acute leukemias generally have been disappointing [87], work in the area of MM has shown some degree of efficacy but major issues concerning toxicity have arisen. Following favorable results in early phase trials [92, 93], pembrolizumab/dexamethasone was included in two phase III trials with and without the immunomodulators lenalidomide (NCT02579863—KEYNOTE-185; NDMM) and pomalidomide (NCT02576977—KEYNOTE-183; RRMM). However, in July 2017, the FDA placed clinical holds on both trials due to the higher risk of death in the cohorts receiving the PD-1 blocker [94]. This suspension soon was expanded to include partial or full holds on all myeloma trials using combinations of immunomodulators and checkpoint inhibitors, an action that later was reversed in the case of three myeloma trials that employed nivolumab (NCT03023527, NCT01592370, and NCT02612779); however, no data have been forthcoming for these resumed studies. At this point, any future role that checkpoint inhibitor-immunomodulator combinations may play in MM therapy is very much in a state of flux [95]. Other checkpoints that may serve as targets for MM but for which only limited preclinical or clinical studies are currently available include killer-cell immunoglobulin-like receptors (KIR) [96], CD47 [97], LAG3 [98], TIGIT [99, 100], and TIM-3 [101]. **Table 1** contains a partial list of checkpoint inhibitors currently in clinical trials for MM.

7. Additional mAbs and their targets

CD138 (syndecan-1), which is overexpressed in MM [107], is the target of indatuximab ravtansine (BT-062), an ADC whose anti-CD138 mAb is linked to a cytotoxic microtubule destabilizing maytansinoid. This agent has been studied in RRMM both in combination with immunomodulators (NCT01638936) and as a single agent (NCT01001442 and NCT00723359). The 34 patients (median 5 prior therapies) in the monotherapy study who received a multi-dose regimen showed a median PFS of three months and median overall survival (OS) of 26.7 months while diarrhea and fatigue were the most commonly reported adverse events [108]. In addition, an anti-CD138 mAb, known as mAb 1610, has shown some potential anti-myeloma promise in a preclinical study [109].

CD74 plays a key role as a chaperone, enabling the proper folding and trafficking of MHC Class II proteins in antigen-presenting cells. In addition, this type II transmembrane protein activates the NFκB signaling pathway following the binding of its intracellular domain to macrophage migration inhibitory factor (MIF) and translocation to the nucleus where it induces proliferation and survival, especially in B-cells. Elevated expression of CD74 in B cell malignancies, such as non-Hodgkin's lymphomas and MM, has made this an attractive target for these types of cancer [110]. STRO-001, which has received Orphan Drug status from the FDA, is an anti-CD74 ADC in which an aglycosylated human IgG1 antibody is conjugated to a maytansinoid linker-warhead. A phase I trial (NCT03424603) of STRO-001 in B-cell malignancies, including MM, recently was initiated [111]. Two other CD74-targeting agents, milatuzumab and its doxorubicin-linked ADC, that had been under study in MM, both have been dropped from further consideration.

Another conjugate linked to a maytansine derivative, the anti-CD56 ADC lorvotuzumab mertansine (IMGN901; BB-10901), had been the focus of a phase I trial in CD56-positive RRMM patients (NCT00346255) but insufficient efficacy and dose-related toxicity reportedly led to discontinuation of further studies of this agent [112, 113]. Other mAbs that have been dropped from further consideration in MM following demonstration of only modest efficacy and/or unacceptable toxicity in trials include the following (target in parentheses): dacetuzumab and lucatumumab (CD40); F50067 (CXCR-4); AVE1642 and figitumumab (IGF-R1); IPH 2101 (KIR); PAT-SM6 (GRP-78); BI 505 (intercellular adhesion molecule-1, ICAM-1), and siltuximab (IL-6).

8. Conclusion

The number of therapeutic options available to treat MM has witnessed a remarkable upsurge since the turn of the current century. The advent of proteasome inhibitors and immunomodulators, in addition to other small molecules working by additional mechanisms, such as histone deacetylase (HDAC) blockade and nuclear export inhibition, has resulted in a major alteration in the clinical approach to the disease. Over the past half-decade, the introduction of mAbs into the fight against this malignancy has further shifted the landscape of how this disease is treated both in newly diagnosed patients and in the relapsed/refractory setting. Chief among these newer entries are daratumumab and elotuzumab, and more recently the anti-CD38 mAb isatuximab and the bispecific antibody belantamab mafodotin. Although employment of mAbs in combination with small molecule agents, such as bortezomib and lenalidomide, has been of immense value in extending patients' ability to achieve deep and durable remissions, relapse and refractoriness to therapy remain as major obstacles to attainment of a cure. Work on checkpoint inhibitors, which have been employed successfully in several tumor types and had shown early promise in MM, continues to move forward in clinical studies of MM, although tempered by recent setbacks stemming from toxicity concerns when used in combination with immunomodulators.

It is evident that future advances in treating MM will be dependent on gaining even deeper insight into the transformative molecular events leading to the disease. As new biomarkers that drive this unrelenting malignancy are identified, design and discovery of innovative target-based therapeutic approaches that will find their way into clinical practice will be established. The attainments already realized by the advent of mAbs, particularly daratumumab, in recent years offers some prospect for even greater success in the application of mAbs in MM over the coming decade.

Author details

Hanley N. Abramson
Department of Pharmaceutical Sciences, Wayne State University, Detroit, MI, USA

*Address all correspondence to: ac2531@wayne.edu

References

[1] Siegel RL, Miller KD, Jemal A. Cancer statistics, 2020. CA Cancer J Clin. 2020;70(1):7-30. doi:10.3322/caac.21590

[2] Kyle RA, Therneau TM, Rajkumar SV, Offord JR, Larson DR, Plevak MF, et al. A long-term study of prognosis in monoclonal gammopathy of undetermined significance. N Engl J Med. 2002;346(8):564-9. doi:10.1056/NEJMoa01133202

[3] Kyle RA, Remstein ED, Therneau TM, Dispenzieri A, Kurtin PJ, Hodnefield JM, et al. Clinical course and prognosis of smoldering (asymptomatic) multiple myeloma. N Engl J Med. 2007;356(25):2582-90. doi:10.1056/NEJMoa070389

[4] Kumar SK, Callander NS, Hillengass J, Liedtke M, Baljevic M, Campagnaro E, et al. NCCN guidelines insights: multiple myeloma, version 1.2020. J Natl Compr Canc Netw. 2019;17(10):1154-65. doi:10.6004/jnccn.2019.0049

[5] *SEER Cancer Statistics Review (CSR) 1975-2017: Myeloma*; National Cancer Institute: Bethesda, MD, USA, 2020.

[6] Jiao Y, Yi M, Xu L, Chu Q, Yan Y, Luo S, et al. CD38: Targeted therapy in multiple myeloma and therapeutic potential for solid cancers. Expert opinion on investigational drugs. 2020. doi:10.1080/13543784.2020.181425. Online ahead of print.

[7] Deaglio S, Morra M, Mallone R, Ausiello CM, Prager E, Garbarino G, et al. Human CD38 (ADP-ribosyl cyclase) is a counter-receptor of CD31, an Ig superfamily member. J Immunol. 1998;160(1):395-402. doi:http://www.jimmunol.org/content/160/1/395

[8] Aarhus R, Graeff RM, Dickey DM, Walseth TF, Lee HC. ADP-ribosyl cyclase and CD38 catalyze the synthesis of a calcium-mobilizing metabolite from NADP. J Biol Chem. 1995;270(51):30327-33. doi:10.1074/jbc.270.51.30327

[9] Laubach JP, Richardson PG. CD38-targeted immunochemotherapy in refractory multiple myeloma: a new horizon. Clin Cancer Res. 2015;21(12):2660-2. doi:10.1158/1078-0432.ccr-14-3190

[10] van de Donk N. Immunomodulatory effects of CD38-targeting antibodies. Immunol Lett. 2018;199:16-22. doi:10.1016/j.imlet.2018.04.005

[11] McKeage K. Daratumumab: first global approval. Drugs. 2016;76(2):275-81. doi:10.1007/s40265-015-0536-1

[12] Dimopoulos MA, Oriol A, Nahi H, San-Miguel J, Bahlis NJ, Usmani SZ, et al. Daratumumab, lenalidomide, and dexamethasone for multiple myeloma. N Engl J Med. 2016;375(14):1319-31. doi:10.1056/NEJMoa1607751

[13] Palumbo A, Chanan-Khan A, Weisel K, Nooka AK, Masszi T, Beksac M, et al. Daratumumab, bortezomib, and dexamethasone for multiple myeloma. N Engl J Med. 2016;375(8):754-66. doi:10.1056/NEJMoa1606038

[14] Dimopoulos M, Quach H, Mateos MV, Landgren O, Leleu X, Siegel D, et al. Carfilzomib, dexamethasone, and daratumumab versus carfilzomib and dexamethasone for patients with relapsed or refractory multiple myeloma (CANDOR): results from a randomised, multicentre, open-label, phase 3 study. Lancet. 2020;396(10245):186-97. doi:10.1016/s0140-6736(20)30734-0

[15] Nooka AK, Joseph NS, Kaufman JL, Heffner LT, Gupta VA, Gleason C, et al. Clinical efficacy of daratumumab,

pomalidomide, and dexamethasone in patients with relapsed or refractory myeloma: utility of re-treatment with daratumumab among refractory patients. Cancer. 2019;125:2991-3000. doi:10.1002/cncr.32178

[16] Moreau P, Attal M, Hulin C, Arnulf B, Belhadj K, Benboubker L, et al. Bortezomib, thalidomide, and dexamethasone with or without daratumumab before and after autologous stem-cell transplantation for newly diagnosed multiple myeloma (CASSIOPEIA): a randomised, open-label, phase 3 study. Lancet. 2019;394:29-38. doi:10.1016/s0140-6736(19)31240-1

[17] Durie BGM, Kumar SK, Usmani SZ, Nonyane BAS, Ammann EM, Lam A, et al. Daratumumab-lenalidomide-dexamethasone vs standard-of-care regimens: Efficacy in transplant-ineligible untreated myeloma. Am J Hematol. 2020. doi:10.1002/ajh.25963. Online ahead of print.

[18] Mateos MV, Dimopoulos MA, Cavo M, Suzuki K, Jakubowiak A, Knop S, et al. Daratumumab plus bortezomib, melphalan, and prednisone for untreated myeloma. N Engl J Med. 2018;378(6):518-28. doi:10.1056/NEJMoa1714678

[19] Musto P, La Rocca F. Monoclonal antibodies in relapsed/refractory myeloma: updated evidence from clinical trials, real life studies and meta-analyses. Expert Rev Hematol. 2020;13(4):331-49. doi:10.1080/17474086.2020.1740084

[20] Beksac M, Aydin Y, Goker H, Turgut M, Besisik SK, Cagirgan S, et al. Early access program results from Turkey and a literature review on daratumumab monotherapy among heavily pretreated patients with relapsed/refractory myeloma. Clin Lymphoma Myeloma Leuk. 2020;20(8):e474-e84. doi:10.1016/j.clml.2020.02.017

[21] Landgren CO, Chari A, Cohen YC, Spencer A, Voorhees P, Estell JA, et al. Daratumumab monotherapy for patients with intermediate-risk or high-risk smoldering multiple myeloma: a randomized, open-label, multicenter, phase 2 study (CENTAURUS). Leukemia. 2020;34(7):1840-52. doi:10.1038/s41375-020-0718-z

[22] Usmani SZ, Nahi H, Mateos MV, van de Donk N, Chari A, Kaufman JL, et al. Subcutaneous delivery of daratumumab in relapsed or refractory multiple myeloma. Blood. 2019;134(8):668-77. doi:10.1182/blood.2019000667

[23] Dhillon S. Isatuximab: first approval. Drugs. 2020;80(9):905-12. doi:10.1007/s40265-020-01311-1

[24] Attal M, Richardson PG, Rajkumar SV, San-Miguel J, Beksac M, Spicka I, et al. Isatuximab plus pomalidomide and low-dose dexamethasone versus pomalidomide and low-dose dexamethasone in patients with relapsed and refractory multiple myeloma (ICARIA-MM): a randomised, multicentre, open-label, phase 3 study. Lancet. 2019;394(10214):2096-107. doi:10.1016/s0140-6736(19)32556-5

[25] van de Donk N, Richardson PG, Malavasi F. CD38 antibodies in multiple myeloma: back to the future. Blood. 2018;131(1):13-29. doi:10.1182/blood-2017-06-740944

[26] Jiang H, Acharya C, An G, Zhong M, Feng X, Wang L, et al. SAR650984 directly induces multiple myeloma cell death via lysosomal-associated and apoptotic pathways, which is further enhanced by pomalidomide. Leukemia. 2016;30(2):399-408. doi:10.1038/leu.2015.240

[27] Moreno L, Zabaleta A, Alignani D, Ajona D, Lasa M, Maiso P, et al. Critical

analysis on the mechanism of action (MoA) of the anti-CD38 monoclonal antibody isatuximab in multiple myeloma (MM). Blood. 2016;128:2105. doi:org/10.1182/blood.V128.22.2105.2105

[28] Raab MS, Chatterjee M, Goldschmidt H, Agis H, Blau IW, Einsele H, et al. Phase I/IIa study of the human anti-CD38 antibody MOR202 (MOR03087) in relapsed or refractory multiple myeloma. Blood. 2015;126:3035. doi:org/10.1182/blood.V126.23.3035.3035

[29] Raab MS, Chatterjee M, Goldschmidt H, Agis H, Blau I, Einsele H, et al. A phase I/IIa study of the CD38 antibody MOR202 alone and in combination with pomalidomide or lenalidomide in patients with relapsed or refractory multiple myeloma. Blood. 2016;128(22):7. doi:10.1182/blood.V128.22.1152.1152

[30] Raab MS, Engelhardt M, Blank A, Goldschmidt H, Agis H, Blau IW, et al. MOR202, a novel anti-CD38 monoclonal antibody, in patients with relapsed or refractory multiple myeloma: a first-in-human, multicentre, phase 1-2a trial. The Lancet Haematology. 2020;7(5):e381-e94. doi:10.1016/s2352-3026(19)30249-2

[31] Fatholahi M, Valencia M, Mark A, Bi MY, Syed S, Zhang YH, et al. TAK-573, an anti-CD38-targeted attenuated interferon alpha (IFN alpha) fusion protein, showed anti-myeloma tumor responses in combination with standard of care (SOC) agents in multiple myeloma (MM) xenograft tumor models in vivo. Clinical Lymphoma Myeloma & Leukemia. 2019;19(10):E116-E. doi:10.1016/j.clml.2019.09.192

[32] Krishnan AY, Patel KK, Hari P, Jagannath S, Niesvizky R, Silbermann RW, et al. Preliminary results from a phase 1b study of TAK-079, an investigational anti-CD38 monoclonal antibody (mAb) in patients with relapsed/ refractory multiple myeloma (RRMM). Blood. 2019;134. doi:10.1182/blood-2019-128007

[33] Lamb YN. Elotuzumab: a review in relapsed and/or refractory multiple myeloma. Drugs. 2018;78(14):1481-8. doi:10.1007/s40265-018-0969-4

[34] Collins SM, Bakan CE, Swartzel GD, Hofmeister CC, Efebera YA, Kwon H, et al. Elotuzumab directly enhances NK cell cytotoxicity against myeloma via CS1 ligation: evidence for augmented NK cell function complementing ADCC. Cancer Immunol Immunother. 2013;62(12):1841-9. doi:10.1007/s00262-013-1493-8

[35] Lonial S, Dimopoulos M, Palumbo A, White D, Grosicki S, Spicka I, et al. Elotuzumab therapy for relapsed or refractory multiple myeloma. N Engl J Med. 2015;373(7):621-31. doi:10.1056/NEJMoa1505654

[36] Dimopoulos MA, Lonial S, Betts KA, Chen C, Zichlin ML, Brun A, et al. Elotuzumab plus lenalidomide and dexamethasone in relapsed/refractory multiple myeloma: extended 4-year follow-up and analysis of relative progression-free survival from the randomized ELOQUENT-2 trial. Cancer. 2018;124(20):4032-43. doi:10.1002/cncr.31680

[37] Dimopoulos MA, Dytfeld D, Grosicki S, Moreau P, Takezako N, Hori M, et al. Elotuzumab plus pomalidomide and dexamethasone for multiple myeloma. N Engl J Med. 2018;379(19):1811-22. doi:10.1056/NEJMoa1805762

[38] Press release (Bristol Myers Squibb - Nov. 6, 2018): U.S. Food and Drug Administration approves Empliciti® (elotuzumab) plus pomalidomide and dexamethasone, a new immunotherapy combination for certain patients

with relapsed or refractory multiple myeloma.

[39] Jakubowiak A, Offidani M, Pegourie B, De La Rubia J, Garderet L, Laribi K, et al. Randomized phase 2 study: elotuzumab plus bortezomib/dexamethasone vs bortezomib/dexamethasone for relapsed/refractory MM. Blood. 2016;127(23):2833-40. doi:10.1182/blood-2016-01-694604

[40] Shabgah AG, Shariati-Sarabi Z, Tavakkol-Afshari J, Mohammadi M. The role of BAFF and APRIL in rheumatoid arthritis. J Cell Physiol. 2019;234(10):17050-63. doi:10.1002/jcp.28445

[41] Pan J, Sun Y, Zhang N, Li J, Ta F, Wei W, et al. Characteristics of BAFF and APRIL factor expression in multiple myeloma and clinical significance. Oncol Lett. 2017;14(3):2657-62. doi:10.3892/ol.2017.6528

[42] Xu S, Lam KP. Transmembrane activator and CAML interactor (TACI): another potential target for immunotherapy of multiple myeloma? Cancers (Basel). 2020;12(4):1045. doi:10.3390/cancers12041045

[43] Rossi JF, Moreaux J, Hose D, Requirand G, Rose M, Rouille V, et al. Atacicept in relapsed/refractory multiple myeloma or active Waldenstrom's macroglobulinemia: a phase I study. Br J Cancer. 2009;101(7):1051-8. doi:10.1038/sj.bjc.6605241

[44] Raje NS, Moreau P, Terpos E, Benboubker L, Grzasko N, Holstein SA, et al. Phase 2 study of tabalumab, a human anti-B-cell activating factor antibody, with bortezomib and dexamethasone in patients with previously treated multiple myeloma. Br J Haematol. 2017;176(5):783-95. doi:10.1111/bjh.14483

[45] Bensinger W, Raptis A, Berenson J, Spira A, Nooka A, Chaudhry M, et al. Phase 1 study of safety and tolerability of BION-1301 in patients with relapsed or refractory multiple myeloma. J Clin Oncol. 2019;37:8012.

[46] Shah N, Chari A, Scott E, Mezzi K, Usmani SZ. B-cell maturation antigen (BCMA) in multiple myeloma: rationale for targeting and current therapeutic approaches. Leukemia. 2020;34(4):985-1005. doi:10.1038/s41375-020-0734-z

[47] O'Connor BP, Raman VS, Erickson LD, Cook WJ, Weaver LK, Ahonen C, et al. BCMA is essential for the survival of long-lived bone marrow plasma cells. J Exp Med. 2004;199(1):91-8. doi:10.1084/jem.20031330

[48] Seckinger A, Delgado JA, Moser S, Moreno L, Neuber B, Grab A, et al. Target expression, generation, preclinical activity, and pharmacokinetics of the BCMA-T cell bispecific antibody EM801 for multiple myeloma treatment. Cancer Cell. 2017;31(3):396-410. doi:10.1016/j.ccell.2017.02.002

[49] Sanchez E, Li M, Kitto A, Li J, Wang CS, Kirk DT, et al. Serum B-cell maturation antigen is elevated in multiple myeloma and correlates with disease status and survival. Br J Haematol. 2012;158(6):727-38. doi:10.1111/j.1365-2141.2012.09241.x

[50] Mori K, Iida S, Yamane-Ohnuki N, Kanda Y, Kuni-Kamochi R, Nakano R, et al. Non-fucosylated therapeutic antibodies: the next generation of therapeutic antibodies. Cytotechnology. 2007;55(2-3):109-14. doi:10.1007/s10616-007-9103-2

[51] Tai YT, Anderson KC. Targeting B-cell maturation antigen in multiple myeloma. Immunotherapy. 2015;7(11):1187-99. doi:10.2217/imt.15.77

[52] Trudel S, Lendvai N, Popat R, Voorhees P.M., Reeves B, Libby III EN, et al. Deep and durable responses in

patients (pts) with relapsed/refractory multiple myeloma (MM) treated with monotherapy GSK2857916, an antibody drug conjugate against B-cell maturation antigen (BCMA): Preliminary results from part 2 of study BMA117159. Blood. 2017;130:741. doi:org/10.1182/blood.V130. Suppl_1.741.741

[53] Lonial S, Lee HC, Badros A, Trudel S, Nooka AK, Chari A, et al. Belantamab mafodotin for relapsed or refractory multiple myeloma (DREAMM-2): a two-arm, randomised, open-label, phase 2 study. Lancet Oncol. 2020;21(2):207-21. doi:10.1016/s1470-2045(19)30788-0

[54] Donaghy H. Effects of antibody, drug and linker on the preclinical and clinical toxicities of antibody-drug conjugates. mAbs. 2016;8(4):659-71. doi:10.1080/19420862.2016.1156829

[55] Press Release (GlaxoSmithKline - August 6, 2020): FDA approves GSK's BLENREP (belantamab mafodotin-blmf) for the treatment of patients with relapsed or refractory multiple myeloma.

[56] Richardson PG, Biswas S, Holkova B, Jackson N, Netherway T, Sofia Paul S, et al. DREAMM-5 platform trial: Belantamab mafodotin in combination with novel agents in patients with relapsed/refractory multiple myeloma (RRMM). J Clinical Oncology. 2020;38(15):TPS8552. doi:10.1200/JCO.2020.38.15_suppl. TPS8552

[57] Usmani SZ, Terpos E, Janowski W, Quach H, West S, Williams D, et al. DREAMM-9: Phase III study of belantamab mafodotin plus VRd versus VRd alone in transplant-ineligible newly diagnosed multiple myeloma (TI NDMM). J Clinical Oncol. 2020;38(15):TPS8556. doi:10.1200/JCO.2020.38.15_suppl.TPS8556

[58] Tiberghien AC, Levy JN, Masterson LA, Patel NV, Adams LR, Corbett S, et al. Design and synthesis of tesirine, a clinical antibody-drug conjugate pyrrolobenzodiazepine dimer payload. ACS Med Chem Lett. 2016;7(11):983-7. doi:10.1021/acsmedchemlett.6b00062

[59] Kinneer K, Flynn M, Thomas SB, Meekin J, Varkey R, Xiao X, et al. Preclinical assessment of an antibody-PBD conjugate that targets BCMA on multiple myeloma and myeloma progenitor cells. Leukemia. 2019;33(3):766-71. doi:10.1038/s41375-018-0278-7

[60] Lee HC, Raje NS, Landgren O, Upreti VV, Wang J, Avilion AA, et al. Letter: phase 1 study of the anti-BCMA antibody-drug conjugate AMG 224 in patients with relapsed/refractory multiple myeloma. Leukemia. 2020:4. doi:10.1038/s41375-020-0834-9

[61] Hau A, Zhu T, Wang R, Lau M, Li L, Li X, et al. Preclinical development of a BCMA targeting antibody-drug conjugate with novel payload for multiple myeloma therapy. Blood. 2019;134:5623. doi:org/10.1182 blood-2019-132080

[62] Singh RK, Jones RJ, Shirazi F, Hong S, Wang H, Wan J, et al. HDP-101, a novel BCMA-targeted antibody conjugated to alpha-amanitin, is active against myeloma with preferential efficacy against preclinical models of deletion 17p. Clinical Lymphoma Myeloma & Leukemia. 2019;19(10):E152-E. doi:10.1016/j.clml.2019.09.253

[63] Nisonoff A, Rivers MM. Recombination of a mixture of univalent antibody fragments of different specificity. Arch Biochem Biophys. 1961;93:460-2. doi:10.1016/0003-9861(61)90296-x

[64] Suurs FV, Lub-de Hooge MN, de Vries EGE, de Groot DJA. A review

of bispecific antibodies and antibody constructs in oncology and clinical challenges. Pharmacol Ther. 2019;201:103-19. doi:10.1016/j.pharmthera.2019.04.006

[65] Stieglmaier J, Benjamin J, Nagorsen D. Utilizing the BiTE (bispecific T-cell engager) platform for immunotherapy of cancer. Expert Opin Biol Ther. 2015;15(8):1093-9. doi:10.1517/14712598.2015.1041373

[66] Topp MS, Duell J, Zugmaier G, Attal M, Moreau P, Langer C, et al. Anti-B-cell maturation antigen BiTE molecule AMG 420 induces responses in multiple myeloma. J Clin Oncol. 2020;38(8):775-83. doi:10.1200/jco.19.02657

[67] Kennedy LB, Salama AKS. A review of cancer immunotherapy toxicity. CA Cancer J Clin. 2020;70(2):86-104. doi:10.3322/caac.21596

[68] Cho S-F, Lin L, Xing L, Wen K, Yu T, Wahl J, et al. AMG 701, a half-life extended anti-BCMA BiTE®, potently induces T cell-redirected lysis of human multiple myeloma cells and can be combined with IMiDs to overcome the immunosuppressive bone marrow microenvironment. Clin Lymphoma Myeloma Leuk. 2019;19(10):e54. doi:org/10.1016/j.clml.2019.09.082

[69] Cooper D, Madduri D, Lentzsch S, Jagannath S, Li J, Boyapati A, et al. Safety and preliminary clinical activity of REGN5458, an anti-BCMA x anti-CD3 bispecific antibody, in patients with relapsed/refractory multiple myeloma. Blood. 2019;134:3176. doi:org/10.1182/blood-2019-126818

[70] Girgis S, Shetty S, Jiao T, Amuzie C, Weinstock D, Watson RG, et al. Exploratory pharmacokinetic/pharmacodynamic and tolerability study of BCMAxCD3 in cynomolgus monkeys. Blood. 2016;128:5668. doi:org/10.1182/blood.V128.22.5668.5668

[71] Raje NS, Jakubowiak A, Gasparetto C, Cornell RF, Krupka HI, Navarro D, et al. Safety, clinical activity, pharmacokinetics, and pharmacodynamics from a phase I study of PF-06863135, a B-cell maturation antigen (BCMA)–CD3 bispecific antibody, in patients with relapsed/refractory multiple myeloma (RRMM). Blood. 2019;128:1869. doi:org/10.1182/blood-2019-121805

[72] Costa LJ, Wong SW, Bermúdez A, de la Rubia J, Mateos M-V, Ocio EM, et al. First clinical study of the B-cell maturation antigen (BCMA) 2+1 T cell engager (TCE) CC-93269 in patients (pts) with relapsed/refractory multiple myeloma (RRMM): interim results of a phase 1 multicenter trial. Blood. 2019;134:143. doi:org/10.1182/blood-2019-122895

[73] Trinklein ND, Pham D, Schellenberger U, Buelow B, Boudreau A, Choudhry P, et al. Efficient tumor killing and minimal cytokine release with novel T-cell agonist bispecific antibodies. mAbs. 2019;11(4):639-52. doi:10.1080/19420862.2019.1574521

[74] Nie S, Wang Z, Moscoso-Castro M, D'Souza P, Lei C, Xu J, et al. Biology drives the discovery of bispecific antibodies as innovative therapeutics. Antibody therapeutics. 2020;3(1):18-62. doi:10.1093/abt/tbaa003

[75] Li Z, Li Q, Zhang G, Ma X, Hu X, Ouyang K, et al. A novel bispecific BCMAxCD3 T cell-engaging antibody that treat multiple myeloma (MM) with minimal cytokine secretion. Ann Oncol. 2019;30:V808. doi: 10.1093/annonc/mdz269.038

[76] Law CL, Aaron W, Austin R, Barath M, Callihan E, Evans T, et al. Preclinical and nonclinical

characterization of HPN217: a tri-Specific T cell activating construct (TriTAC) targeting B cell maturation antigen (BCMA) for the treatment of multiple myeloma. Blood. 2018;132:3225. doi:org/10.1182/blood-2018-99-113921

[77] Pillarisetti K, Edavettal S, Mendonça M, Li Y, Tornetta M, Babich A, et al. A T-cell-redirecting bispecific G-protein-coupled receptor class 5 member D x CD3 antibody to treat multiple myeloma. Blood. 2020;135(15):1232-43. doi:10.1182/blood.2019003342

[78] Richter JR, Landgren CO, Kauh JS, Back J, Salhi Y, Reddy V, et al. Phase 1, multicenter, open-label study of single-agent bispecific antibody T-cell engager GBR 1342 in relapsed/refractory multiple myeloma. J Clin Oncol. 2018;36(15):2. doi:10.1200/JCO.2018.36.15_suppl.TPS3132

[79] Cohen AD, Trudel S, Forsberg PA, Fonseca R, Krishnan AY, Spencer A, et al. GO39775: A multicenter phase I trial evaluating the safety, pharmacokinetics, and activity of BFCR4350A, a FcRH5/CD3 T-cell dependent bispecific antibody, in patients with relapsed or refractory multiple myeloma. J Clin Oncol. 2020;38(15 supplement). doi:10.1200/JCO.2020.38.15_suppl.TPS8551

[80] Smyth MJ, Cretney E, Kelly JM, Westwood JA, Street SE, Yagita H, et al. Activation of NK cell cytotoxicity. Mol Immunol. 2005;42(4):501-10. doi:10.1016/j.molimm.2004.07.034

[81] Gantke T, Weichel M, Reusch U, Ellwanger K, Fucek I, Griep R, et al. Trispecific antibodies for selective CD16A-directed NK-cell engagement in multiple myeloma. Blood. 2016;128:4513. doi:org/10.1182/blood.V128.22.4513.4513

[82] Draghi M, Schafer JL, Nelson A, Frye Z, Oliphant A, Haserlat S, et al. Preclinical development of a first-in-class NKp30xBCMA NK cell engager for the treatment of multiple myeloma. Cancer Res. 2019;79:4972. doi:10.1158/1538-7445.AM2019-4972

[83] Watkins-Yoon J, Guzman W, Oliphant A, Haserlat S, Leung A, Chottin C, et al. CTX-8573, an innate-cell engager targeting BCMA, is a highly potent multispecific antibody for the treatment of multiple myeloma. Blood. 2019;134:3182. doi:org/10.1182/blood-2019-128749

[84] Ross T, Reusch U, Wingert S, Haneke T, Klausz K, Otte AK, et al. Preclinical characterization of AFM26, a novel B cell maturation antigen (BCMA)-directed tetravalent bispecific antibody for high affinity retargeting of NK cells against myeloma. J Clin Oncol. 2018;35(15):1927. doi:org/10.1182/blood-2018-99-118970

[85] Esfahani K, Roudaia L, Buhlaiga N, Del Rincon SV, Papneja N, Miller WH, Jr. A review of cancer immunotherapy: from the past, to the present, to the future. Curr Oncol. 2020;27(Suppl 2):S87-S97. doi:10.3747/co.27.5223

[86] Rowshanravan B, Halliday N, Sansom DM. CTLA-4: a moving target in immunotherapy. Blood. 2018;131(1):58-67. doi:10.1182/blood-2017-06-741033

[87] Liao D, Wang M, Liao Y, Li J, Niu T. A review of efficacy and safety of checkpoint inhibitor for the treatment of acute myeloid leukemia. Front Pharmacol. 2019;10:609. doi:10.3389/fphar.2019.00609

[88] Skarbnik AP, Donato ML, Rowley SD, Goy A, Vesole DH, Munshi PN, et al. Preliminary safety and efficacy data for combined checkpoint inhibition with ipilimumab (I) and nivolumab (N) as consolidation following autologous stem cell transplantation (ASCT) for high-risk

hematologic malignancies - Cpit-001 trial. Blood. 2017;130(Supplement 1): 340. doi:org/10.1182/blood.V130.Suppl_1.340.340

[89] Havel JJ, Chowell D, Chan TA. The evolving landscape of biomarkers for checkpoint inhibitor immunotherapy. Nat Rev Cancer. 2019;19(3):133-50. doi:10.1038/s41568-019-0116-x

[90] Vaddepally RK, Kharel P, Pandey R, Garje R, Chandra AB. Review of indications of FDA-approved immune checkpoint inhibitors per NCCN guidelines with the level of evidence. Cancers (Basel). 2020;12(3):738. doi:10.3390/cancers12030738

[91] Boyiadzis MM, Kirkwood JM, Marshall JL, Pritchard CC, Azad NS, Gulley JL. Significance and implications of FDA approval of pembrolizumab for biomarker-defined disease. J Immunother Cancer. 2018;6(1):35. doi:10.1186/s40425-018-0342-x

[92] Mateos M-V, Siegel D, Shah JJ, Reece D, Avigan D, Orlowski R, et al. Pembrolizumab (MK-3475) in combination with lenalidomide and low-dose dexamethasone for relapsed/refractory multiple myeloma (RRMM)/ KEYNOTE-023. Journal for ImmunoTherapy of Cancer. 2015;3:160. doi:10.1186/2051-1426-3-S2-P160

[93] Badros A, Hyjek E, Ma N, Lesokhin A, Dogan A, Rapoport AP, et al. Pembrolizumab, pomalidomide, and low-dose dexamethasone for relapsed/refractory multiple myeloma. Blood. 2017;130(10):1189-97. doi:10.1182/blood-2017-03-775122

[94] Jelinek T, Paiva B, Hajek R. Update on PD-1/PD-L1 inhibitors in multiple myeloma. Front Immunol. 2018;9:2431. doi:10.3389/fimmu.2018.02431

[95] Oriol A. A critical evaluation of pembrolizumab in addition to lenalidomide and dexamethasone for the treatment of multiple myeloma. Expert Rev Hematol. 2020;13(5):435-45. doi:10.1080/17474086.2020.1744432

[96] Khan M, Arooj S, Wang H. NK cell-based immune checkpoint inhibition. Front Immunol. 2020;11:167. doi:10.3389/fimmu.2020.00167

[97] Sun J, Muz B, Alhallak K, Markovic M, Gurley S, Wang Z, et al. Targeting CD47 as a novel immunotherapy for multiple myeloma. Cancers (Basel). 2020;12(2). doi:10.3390/cancers12020305

[98] Lucas F, Pennell M, Huang Y, Benson DM, Efebera YA, Chaudhry M, et al. T-cell transcriptional profiling and immunophenotyping uncover LAG3 as a potential significant target of immune modulation in multiple myeloma. Biol Blood Marrow Transplant. 2019;26(1):7-15. doi:10.1016/j.bbmt.2019.08.009

[99] Harjunpää H, Guillerey C. TIGIT as an emerging immune checkpoint. Clin Exp Immunol. 2020;200(2):108-19. doi:10.1111/cei.13407

[100] Guillerey C, Harjunpää H, Carrié N, Kassem S, Teo T, Miles K, et al. TIGIT immune checkpoint blockade restores CD8(+) T-cell immunity against multiple myeloma. Blood. 2018;132(16):1689-94. doi:10.1182/blood-2018-01-825265

[101] Acharya N, Sabatos-Peyton C, Anderson AC. Tim-3 finds its place in the cancer immunotherapy landscape. J Immunother Cancer. 2020;8(1). doi:10.1136/jitc-2020-000911

[102] Nooka AK, Manteca MVM, Bahlis N, Weisel K, Oriol A, Alonso AA, et al. DREAMM-4: Evaluating safety and clinical activity of belantamab mafodotin in combination with pembrolizumab in patients with relapsed/refractory multiple myeloma (MM). European Hematology

Association EHA25; Virtual Congress, June 12, 2020. https://library.ehaweb.org/eha/2020/eha25th/294872/ajay.k.nooka.dreamm-4.evaluating.safety.and.clinical.activity.of.belantamab.html?f=menu%3D6%2Abrowseby%3D8%2Asortby%3D2%2Amedia%3D3%2Ace_id%3D1766%2Aot_id%3D23231%2Amarker%3D757.

[103] Nishihori T, Kaufman JL, Hoffrnan JE, Blouch K, Pandit S, Butler E, et al. Open-label pilot study of genetically engineered NY-ESO-1 specific T cells (GSK3377794) alone or in combination with pembrolizumab in relapsed and refractory multiple myeloma. Biol Blood Marrow Transplant. 2020;26(3):S269-S70.

[104] Pirogova O, Darskaya E, Porunova V, Davidova E, Kudyasheva O, Bondarenko S, et al. Study of PD-1 inhibitor (nivolumab) with autologous hematopoietic stem cell transplantation in multiple myeloma patients. Bone Marrow Transplant. 2018;53:657-8.

[105] Cho HJ, Costa LJ, Davies FE, Neparidze N, Vij R, Feng YN, et al. Atezolizumab in combination with daratumumab with or without lenalidomide or pomalidomide: a phase Ib study in patients with multiple myeloma. Blood. 2018;132. doi:10.1182/blood-2018-99-114960

[106] Cho HJ, Cole C, Martin TG, Zonder JA, Fay JW, Vij R, et al. A phase Ib study of atezolizumab (atezo) alone or in combination with lenalidomide or pomalidomide and/or daratumumab in patients (pts) with multiple myeloma (MM). J Clin Oncol. 2017;35(15):8053. doi:10.1200/JCO.2017.35.15_suppl.TPS8053

[107] Fooksman DR, Mazumder A, McCarron M. CD138 regulates myeloma cell survival, proliferation, BM engraftment and tumor organization. Blood. 2015;126(23):2974. doi:org/10.1182/blood.V126.23.2974.2974

[108] Jagannath S, Heffner LT, Jr., Ailawadhi S, Munshi NC, Zimmerman TM, Rosenblatt J, et al. Indatuximab ravtansine (BT062) monotherapy in patients with relapsed and/or refractory multiple myeloma. Clin Lymphoma Myeloma Leuk. 2019;19(6):372-80. doi:10.1016/j.clml.2019.02.006

[109] Yu TT, Xing LJ, Lin L, Liu JY, Wen K, Cho SF, et al. An immune based, anti-CD138 targeting antibody for the treatment of multiple myeloma. Blood. 2018;132. doi:10.1182/blood-2018-99-119112

[110] Stein R, Mattes MJ, Cardillo TM, Hansen HJ, Chang CH, Burton J, et al. CD74: a new candidate target for the immunotherapy of B-cell neoplasms. Clin Cancer Res. 2007;13(18 Pt 2):5556s–63s. doi:10.1158/1078-0432.ccr-07-1167

[111] Shah NN, Krishnan AY, Shah ND, Burke JM, Melear JM, Spira AI, et al. A Phase I open-label, safety, pharmacokinetic, and preliminary efficacy study of STRO-001, an anti-CD74 antibody drug conjugate, in patients with advanced B-cell malignancies. Cancer Res. 2019;79(13). doi:10.1158/1538-7445.Am2019-ct104

[112] Ailawadhi S, Kelly KR, Vescio RA, Jagannath S, Wolf J, Gharibo M, et al. A phase I study to assess the safety and pharmacokinetics of single-agent lorvotuzumab mertansine (IMGN901) in patients with relapsed and/or refractory CD-56-positive multiple myeloma. Clin Lymphoma Myeloma Leuk. 2018;19(1):29-34. doi:10.1016/j.clml.2018.08.018

[113] Jelinek T, Hajek R. Monoclonal antibodies - A new era in the treatment of multiple myeloma. Blood Rev. 2016;30(2):101-10. doi:10.1016/j.blre.2015.08.004